Apocalyptic Ecology
in the Graphic Novel

Apocalyptic Ecology in the Graphic Novel

Life and the Environment After Societal Collapse

CLINT JONES

McFarland & Company, Inc., Publishers
Jefferson, North Carolina

This book has undergone peer review.

Library of Congress Cataloguing-in-Publication Data

Names: Jones, Clint, author.
Title: Apocalyptic ecology in the graphic novel : life and the environment after societal collapse / Clint Jones.
Description: Jefferson, North Carolina : McFarland & Company, Inc., Publishers, 2020. | Includes bibliographical references and index.
Identifiers: LCCN 2020007197 | ISBN 9781476668567 (paperback) ∞
ISBN 9781476639703 (ebook)
Subjects: LCSH: Graphic novels—History and criticism. | Environmental degradation in literature. | End of the world in literature.
Classification: LCC PN6714 .J66 2020 | DDC 741.5/9—dc23
LC record available at https://lccn.loc.gov/2020007197

British Library cataloguing data are available

ISBN (print) 978-1-4766-6856-7
ISBN (ebook) 978-1-4766-3970-3

© 2020 Clint Jones. All rights reserved

No part of this book may be reproduced or transmitted in any form or by any means, electronic or mechanical, including photocopying or recording, or by any information storage and retrieval system, without permission in writing from the publisher.

Front cover images by Hizir Kaya (Shutterstock)

Printed in the United States of America

McFarland & Company, Inc., Publishers
Box 611, Jefferson, North Carolina 28640
www.mcfarlandpub.com

What happens when the world goes quiet ... empty?
—Tommy Jepperd, *Sweet Tooth*

Men have a tendency to bury the truth in dark places, to hide it behind armor
thick and strong ... as if its protections could ever stand for long
against the truly inquisitive...
—Puig Vallès, *SnowPiercer: The Explorers*

Everyone thinks they know how to "fix" the world,
but we'd all be a lot better off if some of us just stayed out of the way...
you know?
—Yorick Brown, *Y: The Last Man*

The thing about smart mother fuckers is that sometimes,
they sound like crazy mother fuckers to stupid mother fuckers...
—Abraham Ford, *The Walking Dead*

All that matters is what you do next.... Give yourself the chance to be better.
—Mary, *The Massive*

...bonum est mortis meditari...

Table of Contents

Acknowledgments   ix

Preface   1

1. Whither Dystopia? Why Apocalypse?   9
2. Dysto-Apocalyptic Hope and the Imagination   31
3. Pathogenic Shaped Futures, Part I: Annihilation and *The Walking Dead*   49
4. Pathogenic Shaped Futures, Part II: Reduction and *Y: The Last Man*   75
5. Post-Human Life in a Post-Nuclear Age in *Snowpiercer* and *Sweet Tooth*   101
6. *The Massive* and Life on a Warming Planet   127
7. Environmental Theory in an Apocalyptic Age   143

Chapter Notes   157

Bibliography   173

Index   177

Acknowledgments

A project like this takes the support of a lot of people and inevitably, in spite of my best efforts, I am certain to omit someone who should be acknowledged here and for that I am sincerely sorry. Still, I would like to thank a number of people and organizations for their assistance and guidance in crafting this book, as I know it would not be nearly as good without them. First, I would like to thank my family who continually sacrifice that I may have the time to think and write. But also for their encouragement and love and interest in the project even if it was feigned at times—I still appreciate what you do for me and love you all.

This book would not exist if it had not been for an opportunity granted to me by the philosophy department at the University of Wisconsin–Stevens Point. While there I was given the freedom to teach a completely self-indulgent class centered on the thesis of this book. I am thankful to the members of the department for many reasons, but I am eternally grateful to the students who enrolled in the course and helped me unpack the issues central to the ideas I explore in this book. Their enthusiasm and genuine interest provided me more motivation to undertake this project than anything else.

The library staff across the University of Wisconsin system, and especially the staff at Stevens Point, deserve special recognition for tirelessly tracking down resource material for this project. I am equally indebted to the University of Kentucky where these ideas first took root and I was able to find ample encouragement to pursue them. The staff at the UK library also deserves recognition for assistance in tracking down many resources for this project and providing a great place to work. Additionally, I had the opportunity to work out some of the ideas contained in this book in the presence of some truly great scholars at conferences hosted by the Society for Utopian Studies and the Radical Philosophy Association. I also owe a great deal of gratitude to many of my colleagues for answering emails, fielding questions, and engaging in long conversations about this topic.

I would also like to thank the team at McFarland that helped to bring this book to fruition. Especially Gary Mitchem, who never once faltered in his support of the idea from our chance meeting in Seattle through publication. An early draft of this work was passed by McFarland to two anonymous reviewers whose careful reading have improved this work immensely. I do not know who they are, but I cannot understate my appreciation of the time and energy necessary to work through someone else's project and provide helpful commentary throughout—especially when reviewing a project as long as this one. Finally, in spite of my best efforts and the assistance of many people there are undoubtedly errors and weaknesses sprinkled throughout the text. These are mine, and mine alone, I only hope they are few and far between and of relatively minor importance in the grander scheme.

Preface

The Visible and the Invisible

 The themes of cataclysm, annihilation, and the environment are the primary focal points of this book, but it is important to bind these focal points within a conceptual range that acknowledges both that catastrophes on the scale I am analyzing are unlikely to occur and that even in the event such an occurrence did take place some life would continue. These two points are necessary because it is vital that considerations such as this do not fan the flames of hysterical exaggeration and to avoid the accusation that there is a morbidity to such an analysis that flirts with irresponsibility. I am interested in this topic for several reasons, some personal others professional, and I believe that properly considering worst case scenarios is a valuable methodology for cultivating a deeper understanding of the real risks facing people around the world. More importantly, to me, is that in the contexts of these real risks we are capturing the frighteningly serious nature of what has been and is being done to the environment. Not for sensationalist or propagandist reasons, but for peace of mind, then, do I think taking apocalyptic narratives seriously is a key to re-evaluating, and hopefully shifting, our current positions on the environment *as an area of genuine social concern*.

 The analysis I offer hinges on the idea that many of our possible futures are bleak and, in some very important ways, how we understand those issues most crucial to life in the future depends a great deal on how we understand those issues now. We inhabit a present that is moving rapidly forward, even as we drag our feet and twiddle our thumbs, pretending to be inoculated by technological safeguards and scientific achievements against the storms looming on the horizon. The texts I have chosen to analyze in reference to this lackadaisical relationship between the present and future are graphic novels. Graphic novels are comic books and that, superficially, puts this argument at a disadvantage because comic books are not typically taken seriously either because they are associated with bad art, stupid stories, and guys in tights; or, they are usually understood to be crude, poorly-drawn, semiliterate, cheap, disposable, kiddie fare. However, as Scott McCloud argues, "they don't have to be!"[1] I believe that our collective cultural response to certain environmental crises is shaped by how a post-crisis society is depicted to us in those media that deal intimately with the future of civilization. More than just what brings about the end of the world as we know it, I am interested in why such case studies do not engender more fear than hope in consumers of such entertainment.

Most post-apocalyptic narrative tales operate formulaically along the storyline that we have done something terrible to cause this disaster, we have learned our lesson, now let us overcome it and start anew. This approach is popular, I argue, because our experiences with disasters, even large-scale disasters, is localized and not totalized. But this type of sentiment and approach is precisely why post-apocalyptic dystopian futures tend to be presented on a hopeful note when, in fact, given the realities of society now, hope would be a long time coming. My argument is built upon the examination of several currently popular depictions of post-apocalyptic worlds.

Each story has been selected because the causation driving the apocalyptic event differs, but the environmental factors that lead to and would develop out of the apocalypse are undeniable. Even so, these popular graphic depictions often fail to consistently make the environment a central focus of their narrative; however, unlike their motion picture and television counterparts the environment in graphic novels is present in a way that it is not in other mediums. In fact, the environment is often more than just the backdrop against which the human drama unfolds but is itself the stage upon which the human drama plays out. Thus, each graphic novel is capable of conjuring creative space to think about the problems humans would face.

There is no denying that popular culture at the moment is transfixed by the concept of a looming dystopian future. Television and movies offer no pretense for the overuse of the dystopian trope—it is, quite simply, what the people want. Young adult literature is currently a thriving dystopian venture and its influence is starting to trickle down into children's literature as well; hell, even our superheroes are engaged in Civil War and zombie apocalypses. Video gaming and tabletop gaming are producing new ways to conceive of and experience the dystopian and the apocalyptic. Additionally, many older dystopian conceptions are being revisited and given new life by being reproduced in different media and made palatable to different audiences. I fully believe that my arguments developed in this book could be applied to any of the above genres and produce the same conclusions.

So, why not television shows or blockbuster movies or literature? The simplest answer is that such an examination would be incredibly expansive and cumbersome, moving between picture and word presentations, so, for clarity's sake and a narrower focus, I have chosen graphic novels, as a melding of the two, for analyzing how apocalyptic futures are interpreted. The simpler answer is that television programs and movies, especially, operate within a limited scope, timeframe, and global catastrophes are represented as "big event" disasters which, no matter how deadly and destructive they are, are short-lived and incapable of crushing the human spirit or fundamentally undermining human society. The simple answer is that in those shows and movies dealing with catastrophic futures humans are given far too much credit for preparation, responsiveness, and ingenuity. Worse, the environment is usually treated as an antagonist to humanity throughout the narrative, presented as an entity still to be controlled and dominated through ingenuity and the sheer force of the human spirit, not to mention that, in the limited time most audiovisual media have at their disposal, environmental narratives tend to play out rather quickly, allowing humanity to "bounce back" or, at a minimum, get a foothold in the process of overcoming whatever cataclysm has happened.

In most audiovisual media, it is as if humans could never succumb to their own destructive lifestyles. Paralleling that, I am not focusing on traditional novels because my concern here is how life and the environment are *depicted* for us, as consumers of apocalyptic futures, providing the images of the world that we associate with life in the aftermath of a catastrophic event. And, while one could argue that movies and television shows also depict the world for us, they leave too little to the imagination and their passive engagement with the audience also distracts from the setting *as a character* even when the environment is the antagonist of the story. The apocalyptic environment in movies and television is treated passively as a non-descript place that provides resources for people working to overcome the catastrophe.

Graphic novels, conversely, utilize the environment in a three-fold way. One, it is the location of the apocalypse and is represented throughout the narrative which, in graphic novels, is a more expansive representation than television and movies can often affect in either scope or detail. Two, it is filler, backgrounded into the story in such a way that even when it is not playing a direct role in shaping the lived experience of survivors it is an ever-present concern. Three, in audiovisual media platforms, the expectation of entertainment and the use of special effects obfuscate the underlying, if obvious, message of peril. Similarly, in books, the descriptions may be starker and more harrowing than in their audiovisual counterparts, but they exist in text only and rely on readers' imaginations to conjure up the hellscape of an apocalyptic environment.[2] Scott McCloud, in an interview with Hillary Chute, argues that literary novels create this problem of monotexturality because the reader is expected to conjure the immediate world the characters are in and the author provides a steady stream of information the reader then uses to continue building the world of the story—but not necessarily the world outside the story.[3]

Graphic novels utilize a combination of depiction and description to present and re-present the post-apocalyptic world while requiring a more engaged reader to really bring not just the story, but the world of the story, to life. Comics are about contingent display, both materially and philosophically, weaving together a constitutive grammar that is not merely opposed to the monotextural nature of literature, but actively subverts it. Comics are thus capable of provoking reader participation in the spaces of the novel that are paradoxically full and empty. To the extent that comics's formal proportions put into play what we might think of as the unresolvable interplay of elements of absence and presence, we could understand the gutter space of comics to suggest a psychic order outside the realm of symbolization and, consequently, along the lines of a Lacanian Real.[4] As such, when graphic novels misrepresent apocalyptic futures the results can be paradigm shifting in ways that could actually make negative future outcomes seem less dangerous or, conversely, more desirable.

It is both hard to fathom and to represent the magnitude of devastation that could occur following an apocalyptic event on a global scale.[5] Perhaps more pressing in such situations is the need to sort out social issues and attempt to gain some understanding of the post-apocalyptic morality that will shape and define how people respond to a situation where everything they have ever known is destroyed. To that end, apocalyptic stories have made great strides in creating discourse about moral permissibility and social order. However, as I have been arguing thus far, these considerations must take place in a world that will provide the *context* for those considerations. One of the biggest advantages graphic novels have over their media counterparts is the varieties of methods that can be deployed by artists and storytellers to foster reader participation as well as helping to shape how the reader might reflect on their participation.[6]

The post-apocalyptic world will be far more dangerous and deadlier than most depictions of the post-apocalyptic dare to tangle with partly, I believe, because of a deep socially embedded anthropocentrism that treats humans as the central and most important focus of any post-apocalyptic event so that human needs tend to trump other concerns. Not that these other concerns are not of import, the problem is that they are treated as secondary to, or as a function of, the human experience, as Susanna Hoffman notes in her essay on the nature of disasters.

> Disasters sometimes strike with the sudden impact of an earthquake or nuclear meltdown. At other times they accumulate over long periods of time with the slowness of a drought or toxic exposure. In whatever manner they arrive, abrupt or subtle, disasters are all-encompassing occurrences. In their wake they sweep across every aspect of human life: environmental, biological, and sociocultural. By their very constitution, disasters spring from the nexus where environment, society, and technology come

together—the point where place, people, and human construction of both the material and nonmaterial meet. It is from the interplay of these three planes that disasters emanate, and in their unfolding, they reimplicate every vector of their causal interface.[7]

A Silent Dance of Seen and Unseen

Another reason I have chosen to focus on graphic novels rather than literary novels is because of the power of the image in contemporary society. Novels face a two-fold problem: they attract a limited, often older, audience on the one hand, which is a problem made worse by the fact that engagement with literature outside of work or school is declining and has been doing so steadily for decades. The National Endowment for the Arts most recent study explains the decline from 1982 to 2015 as going from 56.9 to 43.1. These numbers represent the number of adults who read for pleasure using either physical books or electronic access.[8] Imaginatively this is problematic not only because it corrupts our ability to understand the world hypothetically, but reading has been shown to positively affect how empathetic a person is toward others. My argument is that same empathy could be cultivated and extended toward the environment, along with a richer intellectual understanding of the environmental problems facing society today, if people could really *see* what the world *could* look like if we do not change the way we are living—hence, graphic novels.

On the other hand, novels struggle to attract an audience whose imaginative capabilities have not been constantly tamped down or truncated through years of interaction with passive media formats and continual social media use. The Pew Research Center recently made available its research into who is and who is not reading and the results show that non-readers tend to be disproportionately poor, un- or under-educated, non-white and, by a slight margin, male. However, the numbers of those who read novels regularly are not substantially better.[9] Those years of passive media consumption are usually in front of a television where apocalyptic stories often play out with the same dependence on special effects as film and lack genuine viewer engagement.[10] Graphic novels, contrarily, use images in ways that demand our participation because of how they are put together. That is, rather than being a continuous stream of images, already complete, graphic novels not only rely on readers to fill in certain blanks, but just more generally graphic novels demand our participation as there is no life to a comic except what the reader gives it.[11]

The environmental movements of the last thirty or so years have understood the power of the image in raising awareness for their various causes and, to that end, I contend that graphic novels could be a valuable ally in the struggle to re-shape how we think about what we are doing to nature; but, first, to be effective, graphic novels must begin to depict the post-apocalyptic world accurately and correctly.[12] Many graphic novels are targeted at audiences that capture a readership spanning preteens through adults. Many graphic novels begin as comic books and, as comic books, can take years to deliver the entire story. But, because they are published bi-weekly, or more often, monthly, and require very little time to read, unlike their literary counterparts, it is easy to keep a reader invested long term—irrespective of the use of pictures to complement the snappy dialogue.[13]

Finally, I have selected graphic novels over other forms of media because graphic novels are rapidly overtaking the literary world. Since the late 1990s, when Frank Miller's work on *Batman* and *Sin City* altered the landscape of comics as a genre, making them fit more snuggly into an adult conception of literature, comics and graphic novels have continued to grow in popularity across the world.[14] Alongside Miller's success, Neil Gaiman's *Sandman* and Alan Moore's *Watchmen*, along with several other titles at the time, began to address more "dark" and "adult" themes in the pages of comic books.[15] The popularity of graphic novels and comics today is doubtless buoyed by the onslaught of incredibly successful superhero movies and comic book based television shows that have dominated cinematic and television programing for the better part of two decades; yet, comics and graphic novels continue to be the premier venue for superhero content.

Libraries are seeing their highest rates of circulation in the graphic novel section with many libraries offering titles that appeal to children, young adults, and some 64 percent of libraries offer an adult graphic novel section.[16] Given that a lot of today's best culture critiques are found in graphic novels it is not difficult to see why an older generation, raised on golden and silver age comics, would be drawn back to the medium.[17] Even so, comically formatted social critique must perpetually contend with being seen as inferior when compared to more traditional forms of critique which is why I have chosen to thread my environmental critique through graphic novels in particular.[18]

For many people the present is an uncomfortable situation; the constant struggle to make ends meet in an unstable economy, having to daily straddle contentious fault lines of intolerance and exclusion, increasing amounts of vitriol and hostility between social groups and communities, environmental degradation that is dangerously out of control, and, perhaps most insidious, the repeated failures of resistance and socially progressive movements to effect substantial, lasting, positive change. For those same people the future is often too distant to merit concern given the myriad difficulties facing them in the present. However, in post-apocalyptic stories, the overriding concerns of the present are, in relevant ways, erased. Because post-apocalyptic graphic novel storytelling takes place in the immediate aftermath of a total cataclysm many of those pressing social issues are rendered moot while others are opened up for deeper, more meaningful, interrogation. My concern is with what, taken as a whole, one might reasonably glean from interrogating why so many people on both sides of the many social divides see the near future in such stark and bleak terms and yet express hope for such a future to materialize.

My focus throughout this analysis will be on graphic novels that have, for the most part, achieved substantial readerships along with notable literary recognition. Though there is some overlap in the various stories I utilize I believe that there are significant deviations between them to warrant analysis of each one based on its own world and catastrophe. However, it seems necessary at this early point to make clear that in my analysis I will treat each story as though the reader has read them and is familiar with them, so, I will take no precautions to avoid revealing key plot points or narrative details in making my arguments. At the beginning of each chapter I detail, where the stories are on-going, to what extent I am drawing from each story. If you have not read the graphic novels addressed in the following, but you intend to, stop now and return after you have read them. Otherwise, proceed at your own risk.

Spoiler Alert

Paul Klee once said, "Art does not reproduce the visible; rather, it makes visible." My contention is that the graphic novels analyzed in the following chapters are making visible a present that has serious consequences for our immediate and long-term futures. More importantly, they are making visible the incredibly problematic relationship our culture has with the grand epic narrative that is the apocalyptic tale. One of the signs of a declining society—beyond the insights of Edward Gibbon—is the increasing fascination with epic storytelling and, in that context, fantasizing about the end times themselves. Each of the graphic novels selected and analyzed herein meets that criteria and, as a result, creates an intellectual space for reflecting on what we desire, why we desire it, and what those desires say about us individually and collectively. Epics are grand visions of a social totality brought together in spite of itself and, ostensibly, against all odds. The graphic novels selected for analysis all utilize this assumptive format with respect to the survivors—that is, the characters in these graphic novels either begin with the realization, or come to understand, that in spite of everything that may have gone on before the apocalypse we are now in this together for better or worse.

Each graphic novel under consideration in the analysis that follows also draws upon anthropogenic causes for the apocalyptic conditions that emerge. This is important to note because divine rapture and alien invasion might be legitimate beginning points for postulating the end times, but so are solar flares and asteroids. However, humans could not adequately prepare for gods or aliens any more than they could destruction in more mundane natural forms form outer space. Though some of the graphic novels I have chosen have fantastical elements in them those elements are, in important ways, minimized or inconsequential for the analysis of the stories I am offering or, at the very least, do not make the analysis over, or unduly, complicated. In order to make sense of such attempts I will examine several apocalyptic themes in popular graphic novels that deal with various catastrophic occurrences to tease out what it is post-apocalyptic dystopian visions ought to be divining for us as persons concerned about the state of the world—presently and in its possible future incarnations.

1

Whither Dystopia? Why Apocalypse?

The Obvious Truth Is False

When confronted with the possibility of a maximally bleak future most people find solace in the belief that no matter how terrible things get, so long as there are survivors, human civilization, no matter how low it sinks, will be able to once again flourish. This commitment to flourishing irrespective of what the future holds is, in many ways, similar to Slavoj Žižek's Lacanian analysis of the ecological crisis in *Looking Awry*. Human encroachment upon nature, in a complex multiplicity of ways, represents, for Žižek, the ultimate form of the real. This is because of the radical nature of the crisis itself. The crisis is radical not only because of its effective danger, i.e., it is not just that what is at stake is the very survival of humankind, but rather, our most unquestionable presuppositions, the very horizon of our meaning, our everyday understanding of 'nature' as a regular, rhythmic process.[1]

Of primary concern in the popular construction of the post-apocalyptic is that skepticism about impending crises of various kinds engenders a kind of hyperbole when creating the narratives that attempt to investigate life in the aftermath of such crises. As Tom Athanasiou rightly argues, concerning the relationship between ecological crisis and apocalyptic culture, "the skeptic sees the bitter sadness of green culture as evidence only of apocalyptic yearnings and does not interrogate it further. The skeptic takes it as an obvious truth, known by all sophisticates, that green Cassandras need not be taken seriously."[2] Perhaps the skeptic would also say the concerns of this text are just that, an environmental Cassandra, that need not be taken seriously because, if there is to be an apocalyptic future, surely, the apocalypse will just be a bigger than usual, but still manageable, disaster.

That we will overcome the conditions of a post-apocalyptic dystopian future is the premise of most popular conceptions of the post-apocalyptic. That premise, while questionable for a number of reasons, is made on the assumption that whatever else we have to do to overcome the worst-case scenario we will be able to do so because the world, that is, the environment, will remain stable, usable, and hospitable. Essentially, that the world will remain a *livable place*. The reality, however, is that something direly incorrect lurks in such popular presentations and addressing that is of primary importance and, given the rapidly changing aspects of our current situation, it would seem to be an increasing necessity. First, then, I will approach the problem by examining what the environment would come to look like in the aftermath of an apocalyptic event and, second, demonstrate how such an understanding is largely undermined by pop culture examples that minimize the environment, leaving it in the background so to speak, rather than incorporating it as a main feature of post-apocalyptic life.

It is often the case that when people imagine the "worst-case scenario" of a future to come they do so by amplifying one or a few particularly negative aspects of the social status quo and then allow for a lowering of the overall standard quality of life where everything else is concerned. In this way, the future becomes dystopian, but remains recognizable and, in its recognition, comprehensible. This approach allows the dystopia to provide stinging social criticisms of those amplified aspects of society while ignoring the subtleties, complexities, and interconnectedness of nature and human interaction. Yet, when the future takes on the coloring of the post-apocalyptic rather than the merely dystopian, we are often presented with nightmarish visions of desolation, devastation, and the decimation of life, dwindling the world's occupants to a lucky few who happen to have at their disposal a world that is still serviceable—regardless of whatever damages were wrought upon it in the arrival of the apocalypse. The life of the survivors takes on not only the role of social critique, but the hardy promise of the return to civilization made better, that is, more utopian, because of the hard lessons learned from living through an apocalyptic event—or, more specifically, a human caused apocalyptic event.

However, this very approach renders most post-apocalyptic projects conceptually null because the assumption of an inhabitable world where only a few survivors exists is difficult to maintain coherently. In the event of an apocalypse that is not divine rapture or alien abduction the survivors of whatever calamitous event befalls the earth would find themselves inhabiting a world that would be far more inhospitable than most attempts at depicting such a world have explored. Here I am following Claire Curtis on the distinction between religious interpretations of apocalypse and its more popular culture conceptions as she outlines them in her book *Postapocalyptic Fiction and the Social Contract*. However, while her focus is on the social mine is on the environmental precisely because I think the reality of post-apocalyptic conditions in popular culture are obscured by the intense scrutiny of and focus on the social.

As Curtis argues, "Postapocalyptic fiction, as with critical utopia and dystopia, criticizes where we are now and who and what we might wish to be. Postapocalyptic fiction *reconfigures the conditions under which humans live and demands that humans rethink their premises for peaceful living together. Postapocalyptic fiction moves humans from the state of nature through the social contract and to a new civil society.*"[3] Up to a certain point, I could not agree more. But, before it is reasonable to talk about the reconfiguration of society the context of that reconfiguration must first be considered. And, as the site of the post-apocalyptic, the context which should be given the greatest attention is the environment which, in most post-apocalyptic narratives, is never fully developed, nor accurately portrayed.

The primary difference between my approach and those that begin where Curtis's does is that the re-configuration we ought to be concerned with must take place *somewhere*. If the "somewhere" is fundamentally altered, then our attempts to theorize about social re-configuration must also be proportionally altered. In order to give pause to those who would use post-apocalyptic fiction to criticize how we are living or justify their misplaced hope in a catastrophically shaped future, the environment ought to be thoroughly analyzed and developed as a serious story element *and* as the site of human activity.

Reconciling Dreams and Reality

From the dawn of human life to the age of atomic fission, human craft has increasingly imperiled both environment and habitat. Once social scientific disaster researchers began to realize that catastrophes could be neither understood nor mitigated merely by exploring the physical platform of human existence, acknowledging that social factors were equally relevant, their focus has slowly shifted away from the environment as primary. Instead, subsistence methods, use of resources, construction of shelter, invention of tools, dictates of social structure, distribution of power, attachment to place, mores, and many other sociocultural elements were entangled within the vortex of large scale catastrophes.[4] The consideration of social factors in apocalyptic narratives across the entirety of the apocalyptic experience ought to function as a way to interrogate social norms as causation because large scale disasters involve diachronicity.

Diachronicity is a necessary component of apocalyptic study because crises can only be fully understood by examining them not just before, during, and after the crisis event, but also in tandem with the immediacy of the onset of crisis conditions. Calamities emanate just as much from processes that develop over long periods of time as much as from the sudden onset of crisis conditions. Conditions that spawn, or eventually terminate, in circumstances often viewed as dire, harmful, or horrific usually accrue manifestly or latently under our collective noses. The processes involved range from people's adaptations to their physical underpinnings, to the human manipulation and elaboration of physical surroundings, to the construction of sociocultural institutions, beliefs, and ethos.[5]

Of course, apocalyptic conditions would quite likely not develop in the same ways that large scale disasters might, but that makes it all the more important that in rethinking society through an apocalyptic lens the environment be treated apocalyptically rather than as if a larger than usual disaster had occurred. That is, an event that apocalyptically destroys our collective social structures such that human society could be completely re-thought and re-wrought would also apocalyptically destroy the environment so that any social rethinking would have to be cognizant of what the new world was before society could be refashioned in it. More importantly, the use of apocalyptic tropes to develop social critiques often gives rise to a societal response that welcomes the apocalyptic as a means for gaining access to the opportunity to re-create society. As I will argue in the chapters that follow, each post-apocalyptic world is imagined with a better future in mind *before* the apocalyptic event occurs. Post-apocalyptic stories are hopeful dystopian stories that use apocalyptic events as the eraser which returns us to a state of nature *so that* we can re-fashion the social contract.

The Infamous Bon Mot

A well-known leftist bon mot claims it is easier to conceive of the end of the world than the end of capitalism,[6] and yet, I would argue that such an enterprise is possible only because we have not taught ourselves to fully appreciate the world post-apocalyptically. The end of capitalism would, no doubt, be a global disaster that would affect many people and disrupt numerous social institutions, but the end of the world would be, well, the end of the world, and not just as we know it, but *the end* of the world and a disaster for every *thing*. The wisdom encapsulated in and obscured by the triteness of the cynical leftist quip, is that capitalism is so entrenched socially throughout the world that attempting to conceive of a social order different than or beyond capitalism now requires that we speculate about the total destruction of the social order so that we may begin anew. The primary problem with this fantastical approach to social formation, obviously, is that those conjuring up post-capitalist social orders are relying on capitalistically constructed social mores and norms to frame their post-apocalyptic adventures to say nothing of the privilege embodied by the disproportionately western, white, petty bourgeois survivors.

The difficulty in observing social formation is that there is no *tabula rasa* upon which we can watch a people build a social environment. As Susanna Hoffman notes all human social and cultural situations come to the observer's eye well established and deeply rooted in time and custom. Disaster draws us as close to the basic elements of culture and society as can ever be found. Disasters take people back to fundamentals. In their turmoil, disassembly, and reorganization, they expose essential rules of action, operate on the bare bones of behavior at the root assumptions of institutions, and with only the basic framework of socio-cultural organization. They are likely to dissolve superfluous embellishment and dismantle unfounded or casual alliance. They erase the polish of recent development. As a result, disasters offer unique opportunities in which to analyze hypotheses pertaining to the constitution of society and culture.[7]

Dystopian imaginings capture important benchmarks of apocalyptic warnings, but both problematically ignore the environment and its contribution to life in post-apocalyptic worlds. The infamous bon mot ignores the erasure of society so that whatever post-apocalyptic postulations are relied upon they do not reject the socio-cultural construction that made the pre-apocalyptic so oppressive. Hoffman better captures the realities of the post-apocalyptic as it relates to disaster, but that is the limitation of her insight—disasters, even on a large scale, like the Indian Ocean earthquake and subsequent tsunami that crashed into India in 2004 or the Fukushima nuclear meltdown at the Daiichi Nuclear Power Plant in 2011, are still local events that can be managed by the people directly affected with assistance from the United Nations, allied governments, and, in more and more cases recently, a global response. If we understand how people operate in a disaster situation, and we fully appreciate what an apocalyptic scenario would entail, then it ought to be possible to re-think the global order *as it is* without hoping for an apocalyptic event to provide a clean slate.

Everything Else Is Just Window Dressing

First, it will be necessary to establish what the world—i.e., the environment—would look like if human civilization collapsed globally rather than locally. While I fully intend for the arguments herein to be made applicable on a global scale, I will take the United States as my primary example, generally, recognizing that the conditions the United States would find itself in a post-apocalyptic world are unique to it, the parallels and comparisons to other places would be similar enough that extrapolating from the United States to other first world industrial countries would be easy enough. As such, for both brevity, familiarity, and focus, I am specifically limiting myself primarily to graphic novels that use the United States, or first world industrialized settings as their locus. This is additionally beneficial to my analysis because apocalyptic conditions, should they occur, will not only likely be the fault of the advanced, industrialized world, but they will probably manifest there first.

The developing world would not lag far behind given that much of the existing infrastructure outside industrial first-world countries is imperialistically super-imposed on many countries. In many disadvantaged places dependent upon western generosity living conditions are substantially worse than their western industrial counterparts. Hence, while it may be that the collapse of their infrastructure would not create noticeably worse living conditions than those they are currently expected to contend with, the effects of an apocalyptic event would upend their ability to respond to such a crisis event. People in these areas would, thus, be put into immediate peril, forcing them to shoulder the brunt of apocalyptic fallout. The graphic novels I am working with, with a few exceptions, on the whole, reinforce rather than interrogate the privilege of this narrative structure.

Obviously, a major factor determining developing countries ability to adequately respond to disaster events is the aid received from industrialized countries. A global disaster event would rob them of that aid leaving them with substantial economic, political, and humanitarian concerns that could not be addressed. The goal, then, of such an interpretation is to show how popular conceptions of the post-apocalyptic hinder our ability to generate an appropriate relationship to the status quo—especially where nature is concerned. The imperative driving this critique is the need to address a lack of ecological sensitivity in popular culture, and, in redressing that lack, foster a greater sense of eco-cosmopolitanism.[8]

The focus on eco-cosmopolitanism as a necessary component of any imaginative or real discourse about the future has its roots in political ecology and the way that political ecologists have, thus far, attempted to understand disasters in relation to human societies and the natural world where these two things have too long been considered separate instead of integrated. Though disaster studies are treated differently by sociologists, historians, anthropologists, and other disciplines, political ecology provides a clear-eyed approach to disaster studies because "this branch of disaster study investigates the political and economic structures, policies, and forces that influence and shape the human use of the environment, and stresses environmental use and misuse in a way that other sociocultural investigation does not."[9] My usage of political ecology throughout my analysis is in line with this conception of political ecology, but incorporates a more radical environmental philosophical underpinning.[10] Primarily, though I am concerned with the ways that "environmental attitudes and practices are deeply rooted in historical, cultural, religious, and political structures," I am equally concerned with what I take to be the fact that changes in these attitudes and practices will require radical changes to our lifestyles, values, interpersonal relationships, and how we understand our place in nature.[11]

The distinction political ecology has over corresponding viewpoints is that it treats disaster studies not only as a function of the occurrence of natural disasters themselves, but also as "the function of ongoing social orders as they overlie physical environments. The hazards that lead to disaster, natural or technological, emerge directly from human activity upon environments and the intensity of human environmental intervention. Human societies and their environments are considered fundamentally inseparable, engaged in a continuous process of mutual constitution and expression."[12] Perhaps, then, my analysis should be situated somewhere between the purely academic approaches of standard political ecology and social science-fiction as something like a philosophically rooted socio-political science-fiction. I find this effort to be in line with Peter Frase's efforts in *Four Futures* where he confronts a similar definitional or descriptive project, what he calls "social science fiction," and what I am going to call, by extension, socio-philosophical science fiction.[13]

Learn to Imagine What Is Not Permitted to Be Thought

Disaster dystopias can be instructive in furthering the analysis of apocalyptic ecology in two important ways. First, near-future dystopias that incorporate severe environmental decay often display the readiness with which people believe a failing environment is something that can be overcome no matter how bad it gets. Second, where the environmental destruction is so complete as to render the world inhospitable we often get striking, if limited, access to how dystopian and apocalyptic narratives bleed into one another. That is, if certain dystopian visions fully developed the real conditions of the present *in the terms of* the emerging or established dystopia being imagined, then those stories would become truly apocalyptic.

Consider but two examples of near-future dystopias that incorporate severe environmental decay but fail to produce an apocalyptic narrative, Greg Rucka's *Lazarus*[14] and Rick Remender's *Tokyo Ghost*.[15] Both stories depict near-future worlds lying in ruin, but neither has reached the level of the apocalyptic, and not only because neither of them features massive human population death on par with the graphic novels examined later. Instead, both visions of the world develop particular, currently existing social ills and push them to extremes to produce a future most people would like to avoid. However, far from creating an apocalyptic world, filled with unforeseen horrors that must be navigated anew, the societies that emerge are prepared to capitalize on the disaster because at least some members of society—the social elite—were in a position to absorb and deflect the worst fallout from global collapse. Essentially, these stories develop the future possibility of disaster capitalism on a global rather than local scale.

The local level deployment of disaster capitalism is developed by Naomi Klein in her 2007 book *The Shock Doctrine: The Rise of Disaster Capitalism* and lays out the strategies utilized by corporations and governments that exploit disasters for gains in power before the victims of the disasters can re-stake their claim to what is theirs or prevent the increased exploitation of their lives in the circumstances that develop *after* the disaster. Developing Klein's argument Anthony Loewenstein contends that these ideological maneuvers are implemented "by force, despite the routine opposition to them expressed by populations across the world.... Resistance occurs because inefficiency, abuse, corruption, and death cloud the sunny rhetoric offered by," the agents carrying out the directives of capitalists bent on taking advantage of disastrous conditions.[16] Neither of these thinkers applies their arguments to a global catastrophe, but both Rucka and Remender do, intentionally or not.

One way that humans have historically managed to adapt to disasters is by absorbing or deflecting their effects, either by controlling for the problem—flood plains, dams, etc.—or by developing technologies that respond to certain catastrophic threats—antiseismic or hurricane proof architecture—but the ability of a society to effectively handle the challenges of a disaster depend largely on factors outside the disaster area. These components are usually unaffected by the disaster event, such as potable water and food trucked in from remote locations or government aid for handling displaced populations including the rebuilding of an area.

However, a full scale apocalyptic global event eliminates such resources and so the complexity of an apocalyptic event is capable of rendering useless local precautions and responses to thwart disasters at a micro level and simultaneously worsening conditions at a macro level. Part of the problem is that human technology, designed to prevent or overcome disasters, is a double-edged sword. While enhancing security in some domains, such as hurricane prediction and antiseismic engineering, it also promotes human vulnerability to calamity.[17] In both *Lazarus* and *Tokyo Ghost* it is clear that existing pre-disaster technology and social infrastructure are what allow post-disaster society to continue apace in spite of the hellscape that exists as the background of the storylines.

Rucka's *Lazarus* presents a world that has been crippled by economic inequalities that have rendered governments moot and corporations servants of the powerful families that control the remaining infrastructure and resources. Rucka and his creative team opted for this storyline after thinking about the Occupy Movement and what would happen if the 1 percent lost control, if the economy tanked, and there was no saving ourselves from the fallout. Rucka says *Lazarus* "came out of the Occupy Movement [looking at] the economics of it and positing, 'What happens if it goes horribly wrong? [What if] there is no return from the brink, where are we in 50 or 100 years?'"[18] The resulting story posits sixteen family hegemonies that have divvied up the world's landmass, resources, and population. Each family has authoritarian control over their domain and the people who continue to inhabit the numerous provinces under their control. The non-family member citizens are divided into "waste" and "serfs" where the former have no value and must be cared for minimally while the latter provide crucial services to the families based on talents, skills, or attributes they possess.

Though the world is largely wasteland, especially urban centers of little or no strategic value, the families continue doing cutting edge research, operating educational facilities, developing new technologies and improving old ones, and operating large agricultural, military, and public works ventures. Like any good dystopian story there is constant surveillance and most citizens—waste and serfs—have been microchipped and catalogued in the families' repository of information making anyone traceable, recognizable, and subservient. Even as the world is presented as a negative vision of the near future, and though many people were killed in the economic breakdown of society, it is difficult to assess the condition of the environment because the families have such amazing resources at their disposal for handling the fallout of global economic collapse.

Some areas, especially those where family members reside, are immaculate, clean, and well provisioned, little islands of utopia amidst the flotsam and jetsam of social breakdown. Other areas, such as Montana and the upper West, are relatively stable places, they are inhabited and farmed and utilized by local businesses operating on behalf of family interests, so it is difficult to determine if these places have been envisioned as having survived the fallout of social breakdown or if they have been restored by the people that live there. It is also possible that the families sought to actively preserve these places for the resources they could still provide. What goes unaddressed is how the families managed to navigate the immediate difficulties inherent in social collapse. The result is that *Lazarus* achieves an Orwellian vision of the future that cannot be called properly apocalyptic. That is, in spite of the social collapse at the heart of the story, *Lazarus* does not really envision *apocalyptic social collapse* even though there has been a definite re-ordering of the world in the context of a devastating global calamity.

Remender's *Tokyo Ghost* posits a near future dystopia that emerges from society's addiction to technology. More than that, the world of *Tokyo*

Ghost is the result of runaway consumption and a slow slide into the degradation of self-deception and abuse. The story focuses on two locations, the techno-dystopia that is the Isle of Los Angeles and the tech free garden paradise of Japan. The world of *Tokyo Ghost* represents a Huxlian conception of the future where a population placated with drugs, violence, and various carnal distractions is unconcerned with the stark division of life in Los Angeles into the very wealthiest and everyone else. Unlike *Lazarus*, however, everyone in Remender's near-future can plug-in and opt out, so as long as their needs are met at some minimal level, the social world can continue on *in spite of the fact that the environment is utterly destroyed.*

Remender's rendering of the future is not totally far-fetched, even if it is sensationalized, because he has taken our growing dependence on technology to one of its possible logical conclusions. As he explains via the corporatist ruler Mr. Flak, underscoring the circumstances of life in LA, "I didn't create the problem. Robotics do the farming, mining, manufacturing, construction … everything [which left] an unemployed population with a lot of free time. So, I keep them entertained."[19] The entertainment industry is all that remains for job creation and nearly everyone in LA is techno-altered and techno-addicted in order to escape a very harsh reality of our own making. The vision is one of unchecked automated capitalism developing without corresponding socio-economic shifts in how we structure communities and socio-political responsibilities; simply, without moving beyond capitalism at the same pace we automate society we are, on Remender's account, headed for catastrophe.

That there is such disparity between the haves and have-nots is of no concern to Mr. Flak and his elite counterparts because "the horses eat the apples, the flies eat the shit."[20] For the people in Los Angeles, this sentiment beautifully captures the dystopian circumstances of Remender's imagined future because the degradation of the environment makes it increasingly difficult to meet the needs of everyone; this, of course, causes the elites to look elsewhere for their means of satiation. For a city where Mystic Tunnel Falls empties into the Mystic Sewer a garden island paradise is ripe for plundering.

Tokyo stands in stark contrast to LA as the residents of Japan have transformed the remnants of the city into a wild place where people live more simply and, consequently, more in harmony with nature. Of course, the dystopian aspects of Remender's world never achieve apocalyptic status either because, in spite of everything that had to have happened to bring civilization to this point, including the complete destruction of the environment, society survives and continues to function with, as Tokyo makes clear, the possibility of a healthy nature's survival and the means to manage its destruction in the interim.

Capitalism Cannot Solve the Problems It Creates

Contrastingly, where the environmental destruction is so complete the world has become completely inhospitable to life we often get a striking, if limited, glimpse into how dystopian and apocalyptic narratives bleed into one another. The difficulty being that the environmental problems of utmost concern in the aftermath of an apocalyptic event are erased by the dystopian world being placed in a far distant future. Again, two examples will help bear this out more fully. Rick Remender's *Low*[21] and the graphic novel adaptation of Hugh Howey's *Wool*[22] tempt the apocalyptic with narratives that do take seriously the destruction of the environment and how a globally destructive event would fundamentally alter human society. However, because both are operating in a distant future, it is difficult to really assess how the life altering cataclysms of each story shaped the social order in the post-apocalyptic circumstances imagined to have been caused by the destruction. Put differently, it is difficult to determine if the post-apocalyptic is dystopian *because of* the apocalyptic event or if the social order has *just become* dystopian after the fact as capitalist operators have taken advantage of disaster circumstances.

In *Low*, Remender imagines a world sundered by the sun's expansion into a red giant, an expansion that makes life on *terra firma* impossible, sending earth's inhabitants into the depths of the oceans to live in underwater bubble cities. That humans possessed the technology to accomplish such a thing is not surprising given that there are underwater biomes currently in existence for both research and luxury purposes.[23] Presumably, before society gets to the point that the sun's expansion is a serious concern, humans will have long since mastered the technical aspects of living underwater or even, potentially, interstellar travel. Ignoring the astronomic and allied scientific realities of the sun morphing into a red giant, in the event that earth managed to escape the gravitational grasp of the sun life on earth would be particularly hard—even underwater, if water still existed—and getting off the planet would become an immediate area of concern and, unsurprisingly, that is exactly what Remender has his characters attempting to do.

Millions of years into the future, however, is not a very helpful dystopian, never mind apocalyptic, scenario for interrogating life in the balance now. But, taking Remender's premise seriously in a different way, there is a question worth asking about a future that requires people to live on—or under—the sea. Assuming the environment, in its current condition, does not devolve into something far more untenable than what we are facing now, we are still likely to be facing a crisis of our own making. If the world's population continues to increase, then it is very likely seasteading will become a necessity in the near future. There are currently more than 7 billion people on the planet and more than 9 billion are expected by 2050 with numbers topping 11 billion by 2100.[24] Of course, there are concerns about numbers that high, but the important aspect in determining how those population levels might affect the ability of the planet to sustain us has more to do with our levels of consumption than with the actual number of people on the planet, tentatively speaking. Vivian Cumming, following the lead of David Satterthwaite, argues that a world with a human population of 11 billion might put comparatively little extra strain on our planet's resources.[25] Even if that were the case, overcrowding would definitely be a concern and the race to capitalize on living on or under the high seas would cause a lot of strain on governmental and societal institutions.[26]

There are already developers looking to corner the market for living undersea and there have been fictionalized accounts of seasteading that try to make sense out of the prospects for communal life at sea beyond what a ship, or fleet of ships, might provide.[27] The race to abandon shore could entail a lot of possibilities from small flotilla cities to micronations that could redefine life on an overcrowded planet. Seasteading need not merely be the potential outcome of overcrowding, however, as the oceans could become refuges for many displaced persons in the event of terrestrial cataclysms. Still, life outside of territorial waters would be freed from regulation, allowing innovations to happen quickly, economics could be bolstered by allowing activities existing states find objectionable, not to mention a general piratical anarchism that is likely to develop out of such a social experiment.[28]

Wool, similarly, is set in a distant future, though not nearly as distant as Remender's *Low*, and comes as close to an apocalyptic tale as a dystopian vision can come without making the leap. To be fair, though, *Wool* does present a completely destroyed world inhabited by a few survivors packed into silos dug 144 stories deep into the earth. The residents of the silos are largely unaware of their co-existing counterparts in other silos which suggests a fair amount of time has passed since the silos were dug, built, and occupied in the post-apocalypse. Life inside the silo is largely uninteresting because it is, for all intents and purposes, what is expected of a society living in an overly large tube. What makes *Wool* important for an analysis of apocalyptic ecology is that what keeps the residents of the silo indoors is the toxification of the air which can kill a person exposed to it rather quickly.

One of the aspects of apocalyptic dystopian literature that never seems to get fully developed as a central aspect of post-apocalyptic life is toxicity with respect to vital resources—air, water, and soil—all of which could make living in the post-apocalyptic, or the dystopian, future impossible regardless of whatever else was going on simultaneously. With no indication of what causes the air to be so incredibly toxic one can only reason that it is an accumulation of toxic pollutants which the Environmental Protection Agency describes as poisonous substances in the air that come from natural sources, i.e., radon gas coming up from the ground, or from manmade sources, i.e., chemical compounds given off by factory smokestacks or automobile exhaust fumes.[29] Presumably, the air in Howey's story is something on par with the Donora smog cloud incident in Pennsylvania (USA) that seriously sickened more than 7,000 people, killing at least 20.

In late October 1948, an air inversion caused industrial smog to be trapped low to the ground and prevented it from dissipating into the atmosphere. The resulting cloud of smoke and fog blanketed the community of Donora making it difficult to see even at close range and the elderly and already ill residents of Donora soon began to see major deteriorations in their health. The toll on the community was immense and the event sparked a national campaign for clean air and industry accountability.[30] The clean air fight is far from over as industries continue to pollute the air and fossil fuels continue to dominate energy, transportation, and allied industries. The World Health Organization recently concluded studies that indicate as many as 1 in 4 deaths of children under five are the result of pollution and more than 1.7 million children die every year from environmental pollutants linked to air, water, and unsafe sanitation practices.[31]

Though a proper apocalyptic event would most likely entail a much greater amount of harm than a dystopian vision incorporates, generally, the toxification of a necessary component for on-going life would definitely qualify. And, at the rate air pollution is happening, a problem exacerbated by deforestation, poisonous air is the most likely candidate for toxicity. It is also, however, scenarios like *Wool* that make scientific solutions to apocalyptic problems seem like reasonable responses—science and technology might be able to scrub the air clean, but only if there is a social order in place to utilize those things properly and to maximum effectiveness. Though Howey quarantines his survivors in silos and depicts the world as an inhospitable wasteland, he also preserves social order and, though it is bottled up, life continues largely unchanged. Hence, there is a dystopian vision to *Wool*, like its counterparts analyzed above, that fails to become apocalyptic.[32]

In spite of their many parallels it is important to bear in mind that disaster dystopias, unlike their apocalyptic counterparts, often begin with an axiomatic assumption that a great many more people survive catastrophes than are killed by them, and in the settling dust, among the debris and devastation, society's recovery in the aftermath of a major disaster constitutes the Janus face of catastrophes, the social countenance of perseverance laid over the physical reality of tenuous existence. Post-disaster existences can be a time of not just material but social devastation, fragmentation, and despair and, locally, can achieve something akin to the apocalyptic as I am using it. For many people, survivors and their unaffected allies alike, it can also be, remarkably, a time of social cohesion, purpose, and almost glory. While this seems uncontroversially true, the immediate problem is that in the event of a global apocalyptic incident many more people would die than survive and, while the Janus face alluded to would still exist, the biggest problem with our pre-apocalyptic conditions is that our cultural imagination has fixated on the glorious opportunity of an apocalyptic future.[33]

We Won't Have the Luxury of Digesting One Tragedy Before Another Begins

Long before humans could begin the process of rebuilding civilization they would have to learn how to live in a world that would be rendered largely inhospitable and not just in the short-term.[34] This is where most of the post-apocalyptic visions—especially the most recent ones—fail to make the case for how we ought to be conceiving of the world without the majority of us still populating it.[35] But to get a clearer picture of how the world might shape up in the event of a debilitating global catastrophe it is not enough to treat particularly threatening aftereffects as though they are singular in nature—a charge I would levy against most, if not all, post-apocalyptic conceptions. Instead, as with all things environmental, there must be a concession to synergy, and such a concession means that understanding the fate of humans, human civilizations, and life in general, entails that the world be conceived as a totality that would be, in large part, working at every turn against humanity's remnants. Synergy here does not just mean that events are interconnected or interdependent, but more specifically and more accurately, that events are continuous and contiguous, on-going even when they are not present to our senses.

Though all the aforementioned concerns are not fully worked out in totality in any graphic representations of the end times, they are often present in various attempts at representing dystopian possibilities in many popular culture works, works which nevertheless fail to accurately capture the realities that would be facing humankind in the event of a global cataclysmic event. This is because many writers of end times narratives that frame their worlds as some sort of dystopia rely heavily on the concept that humans and some remnants of human society would survive the end of the world as we know it. In fact, many times these stories depend on the ability of some people, usually members of a well-heeled upper class and a few lowly bottom feeders that happen to be in the right place at the right time, and some human social institutions, to not only weather whatever catastrophe befalls us in the short term, but to make it possible for us to weather the catastrophe long term. Because of this narrative structure many dystopian narratives can present an awful near-future social order and world without having to acknowledge the realities that would follow us into an apocalyptic future simply because a breakdown in contemporary social order is not equivalent to the destruction of human civilization on a global scale even when the breakdown is totalizing.

Hence, to get a clearer picture of the realities facing us given what we have done to the environment, both directly and indirectly, it is necessary to not only separate the apocalyptic from the Apocalyptic, but to understand the apocalyptic social impulse that runs parallel to our interest in dystopian imaginings.[36] There is no doubt that apocalyptic stories have a long and rich history in human society. The familiar Apocalypse is usually cast as divine retribution or rapture, and the Apocalypse is always, somehow, a necessity on god's part to re-create the world *as it should be*.

Certainly, recreating the world in the post-apocalypse is a critical feature of any apocalyptic tale if the apocalyptic tale is going to have any entertainment value. Apocalyptic scale scenarios are the ultimate "what-if" questions—one whose answers can often prove fatal to our ability to really grapple with the consequences of the social order as it has been constructed. Disabusing ourselves of the notion that an apocalyptic event is desirable because it would allow us to re-fashion a better world most likely begins in the simple acknowledgment that our collective socio-cultural mythological re-building of the world after an Apocalyptic event is only possible with divine intervention. The potential for post-apocalyptic recovery, without god's help, ought to make apocalyptic conceptions about our present situation more unsettling than it currently is in our pop culture representations of it.

Many cultures have Apocalyptic narratives in their histories—prehistorical and ancient though they may be. Obviously, the most prevalent today would be the Judeo-Christian and Islamic traditions that dominate Western theological and philosophical metaphysical discourses. It is important to keep in mind that the Biblical apocalyptic stories are not unique to the monotheistic faiths they are located in. There are several religious Apocalypses that fit the mold of the readily understood apocalyptic model most people are familiar with in popular culture, but, given my focus on the western tradition, I will limit myself to the familiar Judeo-Christian Apocalyptic tradition. So, briefly detouring to separate the Biblical notion of Apocalypse from its secular counterpart will prove beneficial as a counterbalance to the reality of an actual secularly caused global apocalyptic event.

Apocalypse: Divine Plan or Human Invention?

An apocalypse is meant to be revealing, and what is revealed is as important as what was keeping it hidden—ultimately, my later arguments are built on a conception of the apocalypse in graphic novels revealing how under-informed and ill-prepared we are for a cataclysmic global event whether it be of our making or not. This is a problem that does not plague Biblical-derived Apocalyptic stories because God, presumably, in destroying the world controls for variable contingencies and synergistic harms. It bears noting, however, that Biblical Apocalypses take place on a world that is substantially underdeveloped making comparisons with more modern imaginings difficult even in a superficial sense.

Biblically, the first Apocalypse most people are familiar with is the "great flood" of Noah and it is arguably the most well-known just generally speaking. However, prior to the flood there was another Apocalyptic event—Adam and Eve's consumption of the forbidden fruit. Eden need not be envisioned as a perfect world, it did, after all, require Adam and Eve to work and keep it.[37] Though Adam and Eve were given dominion it is the case that they were expected to live harmoniously with the natural world.[38] Contained in the Garden of Eden were trees bearing fruit that Adam and Eve were forbidden to consume. Unfortunately, God commands Adam to not eat the fruit of the Tree of Knowledge of Good and Evil *before* he generates Eve from Adam's rib.[39] Subsequently, in her confrontation with the serpent she paraphrases God's command and is, ultimately, tricked into eating the fruit along with Adam. The moment of consumption is also a moment of revelation as "the eyes of both were opened" and in that moment they recognized they were naked and judged that to be a bad thing—prior to eating the forbidden fruit, they are unashamed of their nakedness, afterward, they cover themselves with fig leaves and loincloths.[40]

When God next encounters his beloved transgressors, and learns what they have done, he punishes all involved parties—the serpent, the woman, and the man—the least of which is their banishment from the Garden of Eden. Not only are they sent away, but the way to the Garden of Eden is blocked by ever-watchful cherubim and a flaming sword.[41] Having lost their sanctuary Adam and Eve now toil, and sweat, and live in pain all their days, just to exist day to day and, having been provided garments made of skin by God,[42] it is hard to conceive of a more post-apocalyptic life than the existence of Adam and Eve after their fall from grace. The story of Adam and Eve is *literally* about an apocalyptic event re-making the world. The tranquility of the garden is replaced with the hostility of the new world, the sanctuary of the garden is exchanged for pain, suffering, and loss, the abundance of the garden is replaced with scarcity, labor, and want, the known is forfeited for the unknown and, it bears stressing this point—all because of human desire and consumption.

God, angrily, punishes Adam and Eve for their irresponsibility, but nonetheless provides for them even as they are being banished. Later, when Cain kills his brother, God, again incensed, punishes him as well, though this tale is far less Apocalyptic than his parents' unfortunate experience. After roughly a millennium of toil and strife God's most beloved creations finally reach the point where God, in good faith, can no longer allow humans to exist. The flood narrative related in the Bible is, arguably, the best known annihilative apocalyptic story in the western world. God, seeing that humans had grown increasingly corrupt all over the earth, decides to take decisive punitive action against them. Upon witnessing the wickedness of humans, whose "every intention of the thoughts of his heart was only evil continually," God decides to "blot out man" along with the rest of creation because he regretted having made them.[43]

It is difficult to discern with any certainty what exactly would constitute the level of wickedness that must have dominated humanity to bring God to such a decision since the Bible does not go into gory, gritty detail. However, Jason Aaron uses this ambiguity to develop his re-envisioning of life at the time of Noah in his graphic novel *The Goddamned*.[44] Though Aaron plays a little fast and loose with the Biblical narrative, the pre-flood era of humanity is given a graphic interpretation, and wrestles with the revelation of the Apocalyptic side of God's decision. The world after Eden, in Aaron's retelling, is a paradise that has slowly rotted, a world that "smells of excrement" and is filled with ravagers, rapists, defilers of all sorts who destroy God's creation by their very existence. Of course, it should be noted, that though this world is a dystopian wasteland, it is bountiful enough to provide for the needs of the inhabitants of the world. Noah's elevation to the savior of humankind is called into question as well in Aaron's telling of the story because Noah behaves, in many ways, similarly to the sinners he rebukes on behalf of God, the only difference is his sanctioning by God to bring about a new covenant. As Noah says to Cain, his covenant with the Lord will allow his family to be what Cain's family could not be—worthy of a world all their own.[45]

Biblical Noah found favor with God for alone being righteous and blameless among his generation.[46] God explains to Noah that he intends to "make an end of all flesh" and "destroy them with the earth."[47] Sparing Noah, his family, and two of every living thing, God destroys creation with a great flood when "all the fountains of the great deep burst forth, and the windows of the heavens were opened. And rain fell upon the earth forty days and forty nights."[48] Though the rains last only 40 days it is estimated that the total time Noah and his family are in the Ark is 370 days. This accounts for the rainfall, the duration of the flood, the receding of the floodwaters, and the drying out of the earth.[49] If God's great flood is an annihilative event, then Noah's existence afterward is truly post–Apocalyptic in the strictest sense. And, unsurprisingly, through God's grace, Noah and his family are able to not only survive such a magnificent cataclysmic event, but they are given license to re-populate the earth ostensibly, with a few restrictions, to re-make human society *better than it was before the flood.*

There are, however, similar tales in other cultures. *The Eridu Genesis* in Sumerian cultural creation mythology as well as a Mesopotamian deluge myth in *The Epic of Gilgamesh*. Ancient Greek culture relies heavily on the idea that the world was destroyed periodically by floodwaters, and also fire, which is the portent of the Book of Revelation.[50] Outside Western culture flood myths exist in Chinese, Indian, and Pacific cultures as well as Irish, Mesoamerican, South American, and Native North American cultures. Each is an Apocalyptic narrative unto itself and the erasure of human society is tied up in these narratives as either developing out of the cataclysmic event as a creation story or developing out of the cataclysm as divine retribution and a clean slate to start over. Starting over, with or without divine help, is as at the heart of contemporary popular culture apocalypses.

Other Biblical stories converge on the apocalyptic as well. The Tower of Babel, for instance, though it does not match the level of destruction inherent in the flood narrative, is, nevertheless, an Apocalyptic re-ordering of the world, and Babylon's fall is ultimately linked with the Apocalypse foretold by New Testament Christian eschatology in Chapter 18 of the Book of Revelations. John the Baptist, additionally, is usually cast as an Apocalyptic preacher and there are those that interpret the baptism of Jesus as an Apocalyptic event. Especially where Jesus is thought to be not only a messenger of God, but also the vehicle through which God forgives the sins of humankind *en masse* thereby striking yet a new covenant with humans, similar to his post-flood deal making with Noah, thus re-ordering the world post-crucifixion. It is clear the Apocalyptic tradition has deep roots in human socio-cultural organization. More, apocalypses, as they occur, reveal the wrongs of human society and the post–Apocalyptic is meant to salvage what can be salvaged and then to begin anew. The human love-hate relationship with apocalyptic storytelling is deeply ingrained in our social psyche precisely because of the repetition of the narrative in the Bible. But Biblical Apocalypses are meant to be salvific and the hopefulness thereby engendered is meant to allay fears, not stoke them. As Maggie Nelson correctly notes, "seen in this light, apocalypse seems less of a fear and more of a cheap ticket out of fear."[51]

It is entirely possible that thousands of years into the future, if human civilization is still carrying on, people will have received some sort of apocalyptic tale derived from more modern historical events. The Black Death, for instance, was an order of magnitude greater than anything people had dealt with prior to the decimation of 14th century life.[52] The Middle Ages were rife with end of the world speculations and for many people the Black Death seemed to be it. In addition to the plague, which by itself would have been horrible enough, the world actually got worse in the wake of the Black Death: "charity grew cold, workers grew arrogant, revenues of Church and State dropped, people everywhere were more self-indulgent and frivolous than ever ... the Black Death [revealed] the very extremes of human gullibility and cruelty: the frenzied search for preventives and cures, the depravity of the grave-diggers, the fear of poisoners, and, most horribly, the massacre of Jews."[53] In all reality, that could easily be a description of Biblical Noah's pre-flood world and it succinctly captures Aaron's take on Noah.

Twenty years after the plague had apparently run its course Sir William Petty would write, "Men eat, and drink, and laugh as they used to.... The exchange seems as full of Merchants as formerly; no more Beggars in the Streets, nor executed for Thieves than heretofore," compounding the good fortune of the age by further noting the number and splendor of the coaches, the magnificence of the theater, and the king's possession of a stronger navy and guards than before the "calamities."[54] The closest our world had come to a genuine global apocalyptic event produced exactly the response being two decades removed ought to generate. Given the successful run of pandemics since the plague it is no wonder that our contemporary age is still beset with fears of the microbial. The Black Death surely left its mark upon the collective socio-cultural memory of humanity, but it has been substantially bolstered by outbreaks of cholera, smallpox, Spanish, avian and swine strains of influenza, E. coli, HIV/AIDS, Ebola, tuberculosis, and malaria.[55]

Humans May Survive God's Wrath, but We Likely Won't Survive Our Own Folly

The focus of my analysis is to use popular apocalyptic storytelling to discern how hope for change, for social betterment, became the purview of dystopian visions about the end of the world as we know it. Today the utopian, the person engaged in projects of utopian dreaming, is a modern hybrid. A hybrid in the sense that they occupy two competing versions of their own identity—one rooted in the reality they actually inhabit, day to day, and one fixed, but unanchored, in a future they style as ideal. Much like Le Guin's yin/yang understanding of utopian spaces the utopian fabricator is caught between having to work with what they have and working toward what they desire—they are the crucible that contains the combustible mix of desire and reality.[56] The utopian dreamer must compose their future as being better than the present in terms of beneficence to all—a place where justice, magnanimity, love, and other virtues are amplified, promoted, and made accessible to all the members of society—while simultaneously completely eradicating the many negative aspects of society that permeate the present.

Here we might choose to characterize hybridity in the same sense that Homi Bhabha does in *The Location of Culture*. In Bhabha's interpretation, the hybrid figure opens up a space of cultural uncertainty and instability, an interpersonal space of ambivalence in our cultural identity.[57] For Bhabha, this ambivalent space, or Third Space, "disrupts the unity and homogeneity of cultural identity to create an in-between that can be read anew. The negotiation of identity within the Third Space creates ambivalence at the source of authority and becomes a form of subversion."[58] Ursula le Guin tackles this subversive ambivalence in *The Dispossessed* where ambivalence is rendered as ambiguity and the ambiguity of the utopian existence is meant to tease out the difficulty of navigating the yin/yang existence of the utopian.

The problem faced by the utopian dreamer is not that their utopia is individualized to their imagination or experience. Nor are the problems a product of a disproportionate focus on any particular subjectively important aspects of society rather than the whole as an integrated, overlapping, and often obscure set of issues. The issue is that utopias-as-better-futures represent an ontology of disengagement instead of disentanglement. We may safely assume the subjectivity entailed here does not matter *in the particular* because rational people will identify happiness, health, security, abundance, etc., as the goals of utopia. What those things *mean* to each individual person is a problem of disentanglement; however, most utopias, once they have predicated the dominance of happiness, health, security, abundance, etc., grind to a halt and, thus, falter conceptually because they become static and disengage the dynamism of human experience.

By engaging in an ontology of disengagement utopian projects minimize problems and undermine the ability of people to fully digest the complexity of social issues. The utopia is presented as complete, and in its completeness, its wholeness, it creates a desirable existence without providing a blueprint for how to achieve those utopian features or explaining why they are *desirable*. The utopian assumption is that *if* a utopian feature is recognizable as *desirable*, then it is understood to be so and, as such, the utopian project becomes one of figuring out how to fit all the desirable social puzzle pieces together into a coherent picture. The result is a transgressive individualization that obscures the role of the individual and their complicity in the ills of the social setting they hope to escape by creating utopia. Utopian discourse takes the future as its subject matter and omits the crucial discussion of how we get there.

The obvious problem, then, for an ontology of disengagement is that it disconnects the individual from the reality of what is at stake in seeking a utopian, that is, maximally better, future. This is different than an ontology of disentanglement where the utopian project is one of explaining how to create the conditions for achieving a better future that is built out of the present. Unfortunately, disentangling the various social issues in ways that allow them to be managed effectively to produce the conditions for constructing a better social situation is more difficult than postulating what the goals of social progressivism should be. Hence, the popularity of apocalyptic narratives in popular culture. When the dystopian is presented as the logical outcome of current social practices or foreseen as the natural conclusion to continuing to follow the current trajectory of socially paradigmatic principles, there is no component element that eradicates the entirety of the present status quo. The apocalyptic erases the social status quo in ways that dystopian imagining alone does not and cannot.

The apocalyptic is designed to wash away the entirety of our present social circumstances. The apocalyptic reduces the population significantly, placing the survivors above the pettiness of current social ills—racism, misogyny, classism, etc.—by making their survival dependent upon their ability to rethink their beliefs and moral commitments *with respect to others and other-ness*. More than just reducing interpersonal social concerns to a survivalist microcosm the apocalyptic eliminates supervening social institutions as well so that the foundation of social beliefs can be rethought, redefined, and reshaped allowing for the building blocks of utopia to be formulated. The use of the apocalyptic has taken root as the premier way to re-conceptualize society. Current apocalyptic conceptions of society in popular culture, however, maintain the dystopic negativity usually associated with apocalypse, but, recently, these conceptions have begun to smuggle in the hopefulness of utopian dreaming as the survivors' stories are not just about making it from day to day, but of finding a way to bring civilization back *better than it was*.

More and more people are embracing the dystopian as a form of entertainment, but we would do well to use caution in overlooking the prevalence of dystopian narratives as merely forms of entertainment. The dystopian ideal has long stood for social critique and critical reflection, the unapologetic revelation of what we are in danger of becoming, and then as a form of entertainment. Now, however, the dystopian is starting to shed its skin and has begun to assume the role of hopefulness. I argue, by having done so, the dysto-apocalyptic has made utopianism undesirable and turned the utopian into the standard bearer of disillusionment, cynicism, and apathy—a role reversal of yin and yang. For stimulation of the imagination, action, and hope the average citizen is beginning to appeal to the dystopian prospects of the near future so that the distant future will be a clean canvas. The Apocalyptic is salvific and, therefore, welcome even if it comes from over consuming fossil fuels rather than a forbidden fruit. Because of this it is imperative that we confront the fact that utopian dreaming has become a form of erasure in critical theory and social discourse.

The ideas of utopia have been subsumed by dystopic discourse, the utopian impulse no longer inspires hope to the same extent that it once did, even when utopian thought experiments were employed as social critique. In short, utopia is no longer conceptually adequate for positive social critical theory. As a concept utopia exists under erasure in our current socio-political discourse. The utopian once had a greater ability to command attention to the projects of social betterment they were engaged in. However, the utopian, having drifted away from concrete attempts to realize it, is now more fantasy than science-fiction and more science fiction than reality. As such, the idea of the utopian no longer captures in a comprehensive sense what is at stake in utopian projects for the present political climate. The utopian as a signifier for an individual reduces the individual to an undesirable type—bland, stereotypical, fanciful.

The utopian has become a totalized individual inhabiting a space of non-existence. Often, we do not know if the utopian citizen got that way via a mega-catastrophe or not, but their existence is static and banal especially when compared to individuals experiencing the trauma of calamities. The utopian citizen arrives untested in the imagination, their trials and sacrifices are not on immediate display, unlike the survivor struggling through the dysto-apocalyptic landscapes of total social collapse. Given the realities of mega disasters and large-scale catastrophes, as Hoffman has presented them, the apocalyptic event is clearly a crisis event which must unfold in the human imaginary and in human experience. As Lauren Berlant argues, "crisis is not exceptional to history or consciousness but a process that unfolds in stories about navigating what's overwhelming … an amplification of something in the works [and] in the impasse created by crisis, being treads water; mainly, it does not drown."[59] Perhaps, then, the real problem with apocalyptic hope is that it is rooted in the cultural imaginations belief that we, as a species, have always managed to successfully emerge from catastrophe precisely because we have never wrought a catastrophe which might actually, literally and figuratively, drown us.

2

Dysto-Apocalyptic Hope and the Imagination

Art Should Be an Axe for the Frozen Sea Within Us

The survivor in a dystopian post-apocalypse undergoes a series of responses to their situation that range from the biological to the philosophical and while the recovery process varies considerably from disaster to disaster, by location and type, and from people to people, everyone passes through a number of stages, each of which involves complex acts ranging from the immediacy of meeting needs to searching for explanations.[1] The existential crisis that occurs in the rubble following a catastrophic event allows utopianism to once again be taken seriously. It is the utopia of ruins—part of what makes this desirable is that it distills the ephemeral "we" into an us, that is, "we" the survivors.

The dysto-apocalyptic, when compared to its utopian counterparts, is protagonistically a decidedly collective enterprise and one that, if it is not actual, certainly feels close at hand. In modern dysto-apocalyptic conceptions the survivors are in this together, the problems they face transcend the individual and demand a collective response precisely because society has been decimated, reduced to ruins, and the individual outside the collective is understood to be at risk. In utopias, by contrast, society is an entrenched good and its benefits are not to be challenged or repudiated. Utopias are inert while the dysto-apocalyptic is full of possibility. The result is that it is easy to understand why Winston wants to free himself from Oceania and why Huxley's Brave New World is also undesirable, though both are, ostensibly, utopias. It is also the reason why people desire the opportunity to live through a zombie apocalypse.

If utopia is, as I claim, under erasure, then it will be helpful to take a brief look at how Derrida—from whom I borrow the concept—conceives of the city. Derrida's focus on the city and its role in his philosophical approach to critical projects of social examination treats the city as both a physical and mental space, akin to the Athens of Socrates which existed as the place of his life and as the basis of and foundation for an ideal, *Kallipolis*, which could only be rendered in speech. To concisely state Derrida's position,

> a city ... is at once more than one and less than one, insofar as it subsists only by going outside of itself towards the other, is not merely the houses, the monuments, and the habitat, or to put it otherwise, is not something that can be calculated in terms of its monuments; instead it is the polis where all political decisions are made, a site where all historical events take place and an event that makes the arrival of the other possible. In fact, for Derrida a city embodies the figure of the other.[2]

Utopia so contextualized is, then, the loss of the other, a place wherein all the inhabitants are similar, perhaps not in their individual characteristics, but in their ontology, in their ability to exist as full members of the social order—identical in their citizenry which is constructed and defined in opposition to the non-citizen Other. The utopian project as a deconstructive project is one where the Other is stripped away but social order is maintained. Utopia has, then, become a form of erasure because it has always, even in its most desirable instantiations, been treated as fanciful, as unrealistic, as unattainable—not merely the acceptance of the Other, but the removal of Otherness. The unattainable is a waste of time, resources, and effort. If utopia, then, comes to represent a waste of these things what happens to the ideals attached to utopia?

To Strive for the Impossible Is to Achieve Despair

The utopic future, rooted in love, and comradery, and hope loses its appeal just as it tarnishes these ideals, reducing them to a bland tolerance at best, a passive acceptance at worst. In the rendering of these higher ideals as undesirable the utopian project is reduced to stilted communalism, a rigid cynicism that must be escaped or avoided. Considering Derrida's focus on "cities and how deconstruction is inextricably related to burning, cinders, ashes, ruins, haunting, dissemination and destruction, and at the same time to rebuilding, inheriting, maintaining, opening, reconstructing and welcoming," it is no wonder that Derrida's evocation "of the city as a place of refuge [is] modeled after a certain messianicity, if not messianism,"[3] posits the utopian as a form of erasure. The city is always precariously placed on the edge of its own future as a ghost town, a figment of remembrance celebrated as a site of what once was, and this is always only as far away as the reality that a city-as-place is no longer a place where people want to live. This can come about in a myriad of ways from a rich supply of resources drying up, like so many boom-or-bust gold mine towns, to disease, for example in the aftermath of the Spanish Flu, especially in rural areas, toxicity or, actual erasures through warfare.

2. Dysto-Apocalyptic Hope and the Imagination

The messianic occurs as a product of a set of ideal conditions, the utopian city is a set of ideal conditions, yet, instead of harkening the arrival of the messianic, the utopian city is a ruin, destined to fail and, in its wake, leave only traces of itself that erase themselves from political and historical discourse. Utopias are, quite literally, the invisible that is envisionable, they are the ideal permeated by the problems and social conditions that ultimately undermine them. Oceania, in Orwell's *Nineteen Eighty-Four*, was presented as a utopia, except for Winston, through whom we experience the failings of Utopia—as inertia that intensifies as the utopians start to believe that they have settled the question of "the Good." The Good is central to philosophical ruminations on the ideal society and finds its staunchest champion in Plato who, writing in the *Republic*, argues that mimesis—imitation, or more aptly, art—is a deterrent from the Truth which is why he advocates for banning poets from society. But, it may be argued, that Plato has missed the point, to a certain extent, because art is not always imitative, but rather, reflective. Reflective art, in this sense, is not an artifact that inspires reflection solipsistically as we try to figure out "what does it mean," but rather, it is art that reflects our social condition back at us. It is art that mirrors not art that mimes which makes utopian projects worth considering.

When hope is attached to an unattainable, and undesirable, ideal it becomes seemingly, if not equally, unattainable as well. Why be hopeful for a better future? Why strive for utopia if, given the current circumstances of society, nothing will come of it? Sacrifices become waste. Triumphs become insignificant in the shadow of continued injustice that seems too entrenched to be overcome. The utopian, ever hopeful of a better future, to mitigate the frustration of the unattainable ideal, accepts the reality that we must compromise. Compromise here is not the process of slow, incremental, progressivism, but more akin to the attitude of the voter who, shrugging and sighing, casts their vote for Hillary Clinton and the status quo because the possibility of Donald Trump was unbearable while the future envisioned by Bernie Sanders was unrealistic.[4]

Compromise becomes the metric that weighs one dystopian conception (a worsening of the status quo under Donald Trump) against another (the perpetuation of the status quo under Hillary Clinton) and tries to find a positive, that is, a reason for hopefulness, and finds it in the reality that things could be worse than they are, so, let us preserve and perpetuate the status quo. If Donald Trump turns out to be a genuine harbinger of the apocalyptic, proffering a future that is the total destruction of the status quo, then it is likely his ascension will gradually become acceptable—that is the truly terrifying reality of Trumpian politics in the age of utopian erasure and apocalyptic longing.

Compromise is a faulty strategy because it either gains too little ground toward the ideal or it comes at too great a cost. Either consequence, individually or jointly, becomes the basis for discontent. A condition that is only worsened each time a compromise is struck because every gain is conjoined to a corresponding loss that is also undesirable. Discontentedness can only blossom into disillusion. A self-defeating and self-perpetuating cynicism that bogs down in the ebb and flow of incremental change and inevitable setbacks. Dissatisfaction with the glacially slow pace of improvement is unmitigated by the actual improvements themselves. Hopefulness, then, ends in despair and despair terminates in a desire for destruction of the social order so that a new society can have the opportunity to grow. The utopian prophet may very well seek to bring about the destruction of the status quo, but the utopian is untrustworthy, no longer a marker for hopefulness in the present, but rather a reminder of past failures, and, as such, represents an undesirable means for destroying the status quo.

H.G. Wells was cognizant of this when he wrote *A Modern Utopia* in 1905. He says of his work that the model of utopia must not take shape as a permanent state but as a hopeful stage leading to a long ascent of stages. As envisaged in *A Modern Utopia*,

> Wells's perfect world does not attempt to "change the nature of man." Thus Wells argues that the Morean or Platonic imposition upon individuals of a totalizing schematization—what George Orwell characterizes as the wish to "freeze history"—creates an unsustainably monologistic realm. Wells's Utopia does not negate the past; it embraces and consolidates the dialectical process: "Utopia too must have a history." What remains then is a Utopia that is resolutely material, historical, and dialogical, not the perfected state of fundamentalism but the aspirant condition of a critical humanism.[5]

Disillusionment is the product of frustratingly slow progressive successes where social improvements become not the norm, but elusive. The cycle of compromise wherein every gain is a simultaneous setback creates the conditions for perpetual struggle, Sisyphean in its repetition, until the various goals of social movements become utopian—meaninglessly fanciful—in their very conceptualization and articulation. Paul Ricoeur, following a similar line of reasoning, draws the conclusion that "utopia … is seen to represent a kind of social dream without concern for the real first steps necessary for movement in the direction of a new society [it becomes instead] … a way of escaping the logic of action through a construct outside history."[6]

The utopian ideal is rejected as fervent narcissism, an unrealistic imperialism of the imagination where what is needed is not the utopian, which cannot develop out of the current social context, but rather the dystopian which, unlike its utopian counterpart, always seems to be close at hand—but not just any dystopian formulation of society. The dystopian can be achieved without the creation of the conditions for genuine utopian possibilities—recall

2. Dysto-Apocalyptic Hope and the Imagination 35

Lazarus and *Tokyo Ghost*. Rather, what is needed is the dysto-apocalyptic, the utter destruction of the status quo to make way for the realization of a utopian society brought forth like a phoenix out of the ashes of society. Utopia has become a dangerous idea not because it is unattainable and creates cynicism, but because it ultimately makes dystopia appealing, desirable, and this enables people to actively participate in socially destructive and self-destructive behaviors and to do so while operating under the principle that this type of behavior is necessary in order to achieve the conditions which will allow for a better society to form.

There are obvious consequences of this for Marxist and neo-Marxist projects, and many other critical theories, especially those of Marcuse and Frankfurt School inspired critical projects, whose theories about the future are framed in utopian language. The utopian frameworks employed by theorists through the Cultural Revolutions of the mid and late 20th century are exactly the utopian ideals that have slipped into erasure and a new utopianism, which is developing out of dysto-apocalyptical narratives, is at odds with those philosophies. As will become apparent in the analysis that follows the destruction of the world is becoming a pre-theoretical requirement for developing a narrative that can produce a better society. Breaking with the now familiar bon mot that we can no longer formulate an idea of a future that does not include capitalism because a systematic re-conceptualization of society is only possible in the aftermath of the end of the world, in actuality this means more than just a removal of the military industrial economic structure of society, but quite literally all of the institutions of society—religious, legal, moral, etc. Contrary to Wells, utopia does not *need* a history if it begins in the post-apocalyptic because the apocalyptic is a clean break with history.

The primary concern that develops out of dysto-apocalyptic narratives is that, amidst the collapse of society these narratives present, a stable environment survives that will continue to provide sustenance, shelter, and the necessities required to rebuild civilization. Even when the environment is damaged by whatever circumstances reduce human society to rubble the natural world, that is, *the location of the apocalypse*, is never destroyed beyond recognition despite the many facts and facets of our current industrio-environmental circumstances that suggest this would be the case. How human society is rethought in the aftermath of apocalypse, how the possibilities of post-apocalyptic life are presented, makes these narratives important artifacts for critical theorizing and critical engagement. Using the false pretenses of survival in the post-apocalyptic ought to help strengthen current socio-political discourses about the need to alter the status quo both to move away from apocalyptic narratives, but also to come to terms with the realization that continuing along our current path will, ultimately, lead us to apocalyptic circumstances *which ought not be desirable*.

Apocalypse: A Numbers Game

To begin conceiving of an apocalyptic world it is necessary to establish how much life would still exist and since most post-apocalyptic visions of the future begin with a few scattered survivors let us assume that whatever befalls humanity—for now, assuming the destruction somehow falls squarely on humans and our animal cousins escape largely unscathed—then let us assume the worst: a 99.99 percent die-off. Taking the current global population at 7.22 billion and granting that 99.99 percent of the global population would die the world would find itself populated by a meager and approximate 722,000 people. Of course, tinkering with the numbers within the scope of whatever is chosen as the reason for the destruction of so many people could alter the number significantly, if ultimately insubstantially—for instance, by merely pursuing only a 99 percent die off the global population would be roughly 72.2 million. The justification for using the higher 99.99 percent is the near total destruction of the human race that usually accompanies an apocalyptic vision of the future—something easily achieved by disease, large scale nuclear war, or any number of other factors or combination thereof. Also, there is no need to assume that the deaths are simultaneous to one another—that is, one could easily project that it would take several months to achieve such a vast number of dead. If the scenario in question is genuinely apocalyptic, however, the scale of death necessitates a mortality rate that is both near annihilation and occurs with a rapidity that would outpace an organized, concerted, social response.

Using the higher death toll, consider the impacts on population facing the United States. The current population of the United States is roughly 325 million with a population density of approximately 92 people per square mile. If the United States suffered a population reduction of 99.99 percent that would leave about 33,000 people in the United States spread unevenly across the country. The population density would drop to approximately one person every 108 square miles. Assuming the survivors are bunched together in small collectives and pockets, then the concerns facing those individuals would be two-fold: one, would they have enough people and the right people to maintain the environment in the area they occupied, and two, what would happen to the vast swaths of the country that had too few or no persons occupying it.

Synergistically speaking the greater concern is the second since whatever was happening to the environment outside the sphere of control of the survivors would most certainly spill over into the places they were occupying and, either immediately or in time, completely overwhelm them. The aftermath of an apocalyptic event is literally the worst time to be living downstream. Again, as an example, this is a fairly shallow one, since the United States would likely suffer greater or lesser population losses and higher or lower rates of devastation—depending on the apocalyptic event—than their neighbors or other countries since the apocalyptic circumstances would undoubtedly affect different civilizations and cultures in differing and unique ways. This also includes how capable they are at responding to such an event and First World countries today, thanks to centuries of exploitation, are undeniably better equipped to marshal resources and organize a response, at least in the short term.

Using the highest percentage, however, combined with the realization that the dead would not be evenly distributed—it is easy to accept that the most populous places would suffer the worst, China and India with over a billion people each would likely be driven to emptiness, while smaller populations or isolated ones, island societies, for example, might escape the devastation largely intact—depending, of course, on how high the seas rise in a given set of apocalyptic circumstances. Regardless, one of the biggest concerns driving the truly nightmarish nature of the post-apocalyptic becomes readily clear: there would not be enough people to maintain the very infrastructure we have built to sustain us, the very collapse of which would be our long-term undoing. In its unraveling the fate of humanity and the environment would be interlinked and the potential for life would dwindle to something far more akin to complete annihilation than even the 99.99 percent we began with.

However, granting even the smaller percentage of dead, the world, as we have built it, would largely grind to a halt. A couple of factors that there would be no controlling for would immediately come into play. One, there would be no way of knowing who would survive. Not just on a scale model, but the actual "who," the individuals that would comprise the survivors, and given the highly-specialized composition of most western societies today, it would be instantly disastrous if certain skill sets found their way too quickly into extinction. Yet, again, we know this would likely be what would happen since certain highly skilled labor sets are already in the minority of any given population's common knowledge pool. For instance, how many people can run a nuclear facility? How many people can repair the robotics and technologies that might assist in stabilizing oil refineries or chemical plants? How many people could manufacture medicines? By allowing industry to be guided by principles of profit too few people per capita exist in certain industries to ensure that our own creations would not eventually conspire against us in the struggle for survival in a post-apocalyptic world.

Second, where these people would be located would also be a problem because of the decentralized nature of the infrastructure and the scale of it. How would an engineer in Florida be of any use to people in St. Louis or New York trying to keep their bridges from collapsing? How would farmers in Oklahoma be of any use to people stranded in metropolitan food deserts like Los Angeles? These are examples that are continentally local at least, but if the needs were transoceanic, who would fly the planes or navigate the oceans if all the qualified captains and pilots fell as statistics in the apocalypse? Perhaps more perplexing, and adding to the easily recognized problems that would create the conditions of the post-apocalyptic world, some of the survivors would be worse than unskilled—an unskilled person could contribute in at least some positive way; but some of the survivors might actually be mad, quite literally, sociopaths or worse, for whom a lawless urban world slipping quickly into decay would present a plethora of opportunities to wreak havoc unabated, sowing destruction in their wake unchecked by social strictures that once constrained their behavior, accelerating the process by which certain terrible things might come to pass.

Allowing a reasonable $1/1000$ ratio of destructive, anti-social personalities to survive the apocalypse, that is 722 persons who could be potentially working, intentionally, to make matters worse. There is no controlling for who would survive and even if half that number were in lock-up already, surviving but behind bars, there would still be 361 people running wild who may not care to see the world burn. This number obviously does not account for certain anarchist worldviews or persons holding similar philosophies who might see the cataclysmic as an opportunity and not as a disaster.

It is the composition of the survivors that would matter most, and mostly because, in the immediate aftermath of an apocalyptic event, those skills deemed most important would be the ones that provided food, security, and shelter. Moreover, everyone—in a survivor camp or literally on their own—would be turned to such things by the very necessity of lack and scarcity that would emerge rather quickly in a post-apocalyptic setting. While the few bedraggled yet intrepid survivors were busily securing adequate shelter, food, and supplies, however, the world around them would be failing—not just falling into ruin but transforming into an inhospitable and hostile place. This world, the one that would materialize from the apocalyptic upheaval of society, is the world that we need to be attempting to understand and, to date, that understanding has largely eluded full expression in popular culture giving rise, unfortunately, to a false sense of hope. To better understand the scope of the harrowing nature of the post-apocalyptic landscape it is important to understand certain elements individually, thus allowing a clearer picture of the post-apocalyptic landscape and what it holds in store for those who survive to see it.

Apocalyptic Context

If the world were to meet with calamity in the near future our fossil fuel-based society would conspire rather quickly to hem surviving communities

2. Dysto-Apocalyptic Hope and the Imagination

into relatively small and equally unstable pockets of livable spaces. In particular, extractive industries—especially coal, oil, and gas—would leave behind deadly artifacts that would be environmentally crippling very quickly. While each industry is, currently, committed to convincing people how necessary they are to society they are not terribly keen on ensuring their processes and products would not become terminal problems in an apocalyptic landscape. Yet, anyone surviving into the post-apocalypse would have to contend with the remnants of each one as well as the problems that would exist at the confluence of the destruction left in their respective wakes.

The coal industry has a long and ugly history in the United States and the industry would not stop destroying the earth even if every person involved with the coal industry were taken in the apocalypse. In fact, such a thing would be particularly unfortunate with respect to slurry ponds and empty, increasingly unstable, and difficult to find or trace mineshafts. The coal industry, for example, has forever altered the Appalachian landscape via a coal extraction technique known as mountain top removal. But the problem of mountain top removal is not isolated to the reduction of a mountain to a plateau, rather, it is in the sheer number of backfilled streams, toxic soils, and habitat destruction in and around mountain top removal sites. There are currently 501 active or abandoned mountain top removal sites in the Eastern United States encompassing a staggering 1.16 million acres of land left poisonous, unstable, and desolate—conservative estimates put natural forest reclamation in the hundreds of years.[7] A time period during which the very ecological composition of Appalachia may radically shift given the unmitigated conditions that would exist in coal country. However, this is far from the only problem that would follow us into the post-apocalyptic landscape.

Coal is a dirty industry and, as such, coal requires cleaning, and cleaning coal produces a lot of tonnage of highly toxic water that is usually stored in slurry ponds. Slurry ponds contain high levels of heavy metals, a single spill capable of poisoning even more acres than extraction pollutes. There are "more than 500 slurry ponds in Appalachia and hundreds more around the country," built on the cheap and maintained at minimum government standards.[8] Spills are common even today with some oversight and plenty of people around to raise the alarm, but without enough people, or people skilled enough to predict or prevent spills, it would only be a matter of time before they all let go. How bad would the devastation be to the environment if that came to pass? Consider but one recent example.

On October 11, 2000, a 72-acre slurry pond collapsed into the mine beneath it spilling 300 million gallons of toxic slurry into the water supply of Martin County, Kentucky. The spill saturated local streams and the Big Sandy River ultimately tainting 110 miles of Kentucky and West Virginia waterways.[9] There are more than 500 more of these ticking time bombs, each as deadly and each as poorly monitored and maintained as the Martin County pond, and with few people qualified to deal with a problem, both of its type and scale, between coal mines, removal sites, and slurry ponds, most of Appalachia and the areas downstream would become toxic. Add to this potential devastation Eastern Coal's cousin in the West, deep pit mining, which is, in its own way, mutilating and transforming the Great Plains, and coal's legacy in the post–Apocalypse future will be buried streams, missing mountain and plains habitats, toxic water, soil, and plants, and, without the tools, people, or knowledge of how to clean up the spills, the damage could last many hundreds of years. Sadly, coal is but one toxic extractive industry that uses such models for business-as-usual, add to it copper, gold, and other metals and the United States would become a landscape pock marked by toxic sinks that would fundamentally alter life in the post-apocalyptic world extending hundreds of miles in every direction from each site.

Like coal, oil, too, would leave its mark upon the landscape. The oil reserves stored in refineries around the country would eventually leak out. Houston, alone, could poison the entire western coast of the Gulf of Mexico and much of the delta region where the Rio Grande meets the sea. Even if the refineries were shut down properly it would not take long for the petroleum—versions ranging from sticky crude to refined liquids—to begin to leech out into soils. A couple of well-placed lightning strikes could start a fire or two that would rage, unabated, potentially spreading throughout the Houston-area oil fields, burning day and night, spewing hot, toxic ash into the atmosphere polluting a range from southern Arkansas to northern Mexico with the right winds. But Houston is not alone in its ability to utterly destroy the post-apocalyptic world.

There are 142 operable oil refineries in the United States housing an average of almost 18 million gallons of oil a day.[10] This, of course, does not account for pipelines, off-site storage, or the by-products of production—all of which would also contribute to the oil industry's ability to re-shape the post-apocalyptic landscape. Again, given so few people, and even fewer with the knowledge and skills to manage or avert catastrophes in the oil industry, whole areas of the United States would become inhospitable to human life and, likely, life in general while hundreds and hundreds of years would have to pass to soak up the oil and safely return it to the long-buried strata where we found it. Between coal in the east and west and oil throughout the Midwest, including a potential near total loss of Texas, there are dense collections of refineries all over New England, the Ohio River Valley, and Southern California. Simply assuming enough people were around to maintain several of these facilities, most of them would go to rot over time leaving in their wake a futuristic Black Death to rival the medieval plague.

The current trend in hydraulic fracking is the lesser of three evils but not by much and primarily only because it is relatively new. Fracking has already been linked to instances of earthquakes and natural gas leeching into ground water transforming it into a combustible substance. Besides adding to the toxicity of water, its potential to explode or burn means that water could become a fire hazard which could be a significant problem in areas of heavy forestation especially if severe droughts were to take hold. This is in addition to the instability created by the potential increase of earthquakes. These problems are difficult to manage precisely because they are not containable or localized and the potential consequences of fracking are unpredictable.

Hence, controlling for the problems caused by fracking—especially if the practice continues apace or gains in popularity—is not only difficult, but even if people with the knowledge and skill to do something about it survived, there is not much to be done because there would be so much that needed to be done. Unlike mines or oil wells, and unlike slurry ponds and refineries, fracking is below the surface, its damage happening out of sight, and its effects not manifesting until it is too late. Survivors—even unskilled or unknowledgeable ones—could battle a fire, or patch a dam, at least long enough to organize a hasty retreat to a safer locale, but fracking gets you by surprise, and unlike oil slicks and coal sludge, gas is invisible, so you would not know you were being poisoned until it was killing you. Combine the perils of the gas industry with coal and oil, however, and the post-apocalyptic landscape would very quickly become unstable, highly toxic, and hundreds of years in waiting for natural reclamation *beyond* the recovery time necessary to overcome the apocalyptic event.

The Apocalypse Gets Worse

The legacy of the fossil fuel industry will not be isolated to the physical landscape, unfortunately, because as the motor of global climate shift the effects will continue for some time even if all industry ground to a halt and every car were to sit idle.[11] There would be carbon contributions to the current global warming trend for some time after the apocalyptic event primarily from large scale fires, whether intentionally set off by an errant arsonist or by a chance lightning strike, as well as other natural phenomena, such as the continued thawing of permafrost. In the short term, while the earth and its atmosphere were busy recycling and storing the carbon-based maelstrom we will have unleashed upon it, the effects of global warming would continue to plague whoever managed to make it through the apocalypse.

The immediate consequence of this would be that survivors would have to become nomadic with no idea how far they would have to range in order to care for the individuals in a given social cell. On an ideally healthy planet estimates of the land-to-person ratio for sustaining a nomadic hunter-gatherer group are approximately 40 hectares per person. This number increases rapidly as the ideal slips into apocalypse. In merely unfavorable conditions or marginal environments the ratio of land needed for individual survival can be as high as 10,000 hectares per person. This is just shy of forty square miles per person. In a post-apocalyptic setting that number could easily double or triple and without reliable, efficient transportation, and sans the ability to travel as the crow flies, itinerant survivors would likely face starvation conditions long before the hostility of an inhospitable environment caught up to the them.[12]

Assuming, generously, that the survivors in such a setting were able to meet their needs, increased storm activity would present them with a two-fold problem. On the one hand, there would be, as has already been observed, an increase in the number of storms. On the other hand, there would be an increase in the intensity of each individual storm. This is not merely a seasonal problem either but rather one that is by turns, affected by the seasons, in part, but also by storm-type. Hence, there would be an increase in hurricanes, tornadoes, tsunamis, blizzards, along with dust storms, and heat waves, and deep freezes, but these would not necessarily appear where one would expect. Recent weather patterns harken a far more discouraging trend as hurricanes have not only increased in number and strength, but in where they are appearing, hurricanes recently striking Spain and New England for the first time in recent history. The damage these storms would create along coastlines and in coastal cities would exacerbate already terrible conditions in those locales not to mention spreading the accumulated aftereffects across a broader area than the urban environments they would, at least initially, be isolated in.

New York, for instance, would likely succumb to ground water flooding the subway tunnels, but a heavy blow from a category five summer hurricane could reduce the city to unlivable nearly instantaneously. The secondary side-effect of a city having been rendered unlivable is that in these cases it is not merely that they are abandoned locales, but that the resources they hold for the survivors would be rendered inaccessible. A flooded New York or an ever-burning Houston—or any large population centers with lots of resources—would be impenetrable and all potential resources would be lost, or incredibly difficult to acquire, significantly reducing the ability of the survivors in proximity to those places to increase their chances of survival.

As the planet continued to heat and weather patterns shifted storms would become unpredictable and more frequent making it just as dangerous to live inland as along the coast. More, even without severe drought—though current predictions have the United States consumed by extreme drought conditions as early as 2090—lacking access to air conditioning and ample water, the American Southwest would become a place few could live which means that six of the ten most populous cities in the United States would have to be abandoned as infrastructure failed and the landscape turned against its inhabitants.[13] All of this on its own would be adequate to create quite the dystopian narrative, and yet, the story of storms in the post-apocalypse is made more deadly by the fact that the storms themselves would be responsible for shifting and shuffling the fossil fuel fallout around a very wide area both inland and out to sea.

So, where most dystopian visions might imagine terrible storms rarely do they depict the damage as being frothy with oil or coal sludge in the aftermath of the deluge; nor do they attempt to extend the destruction of tornadoes beyond the path of the storm, but in a world with a substantially depleted population many cities would burn, flooding would follow, and the destruction would leave not just broken buildings in the wake of such occurrences, but places made unusable by people except in very limited, and always dangerous, ways. Heavy snows, without interior heat or people to manage the snowfall would collapse roofs and pull down power lines, which in the earliest days of the post-apocalypse, could still cause fires, but more likely it would act as a catalyst for rot as wires were pulled from buildings exposing the walls to moisture thus, eventually, incubating molds. What we have done to the climate will eventually subside, but the time it would take to accomplish that would see several generations of survivor's children grow old and die—though how old and what would kill them is a debatable feature of the post-apocalyptic landscape.

The Incurable Spread of Fear and Disease

Another problem that would quickly make it difficult to live in the post-apocalyptic world would be the lack of medicine, specifically inoculations against diseases. Ignoring, for the moment, the failed logic of the anti-vaccine movement, anti-vaxxers present a glimpse of what life would be like without the wonders of modern medicine. Absent access to medications and vaccines, which would only last until the currently available supply of them, plunderable in stores, hospitals, doctors' offices, and locatable warehouses, were exhausted, life will quickly revert to one punctuated by epidemics. Soon, various poxes, strains of influenzas, polio, and other communicable illness which, left unchecked become deadly, i.e., syphilis, would be a constant, and constantly increasing, threat. The elderly and the very young would be most at risk, but pandemics would be a single sneeze away as more and more people attempted to flee inhospitable living conditions—imagine what it would be like in the early days of a cataclysm when the American

Southwest started spewing refugees. The six most populous cities housing more than one million people each, Los Angeles boasting almost four million alone, and this is not including surrounding areas and small cities and towns. Those six cities alone would put potentially 10 million people into other areas.

If a microbial or viral pandemic that was capable of extinguishing humanity emerged, then it would definitely wreak apocalyptic havoc, a theme dealt with more fully below, but the concern is nonetheless real. As recently as January 2017 a woman in Nevada (USA) fell victim to a "superbug." The woman's death was reported by the CDC in their *Morbidity and Mortality Weekly Report* of January 13.[14] The woman died of an incurable infection, resistant to all 26 antibiotics available in the U.S., a case indicative of the future which medical professionals refer to as a "nightmare superbug" because this particular specimen was even resistant to antibiotics developed as a last resort against bacterial infection.[15] Though the Nevada case is not the first, and recognizing that there are other contributing factors to why some people succumb to antibiotic resistance infections, it is an area of growing concern given that the CDC estimates 23,000 people die every year in the United States from multidrug-resistant infections and a corresponding British report, *The Review on Antimicrobial Resistance*, estimates that globally, 700,000 people die each year from infections that are drug-resistant.[16] The Black Death of the 14th century may have been a practice run compared to what an unleashed nightmare superbug could do today.

If the die-off was not instantaneous or not significant enough to ensure that massive amounts of refugees were not left moving around the countryside not only would overcrowding put incredible stress on limited and quickly dwindling supplies of food, housing, and medicine, but overcrowding would spread diseases and, as more and more people died, more and more of the dead would include doctors and nurses. As medicine runs low and rationing becomes a priority disease would present yet another problem that would be a fixture in the post-apocalyptic landscape for some time—dead bodies. Bearing in mind that this is only accounting for deaths attributable to the cataclysmic event and its particular aftermath and further assumes that an apocalyptic event would bring communities together for the greater good. A more reasonable rendering of life in the aftermath of an apocalyptic event that did not include an instantaneous die-off would include conflicts over territory, resources, political power, and deeply entrenched socio-cultural differences that would inevitably contribute directly to the body count—that is, on what grounds do we assume Locke was correct and not Hobbes?

Disposing of the dead would rapidly develop into a problem that would require lots of labor power—a resource not being dedicated to maintaining law and order, but also not to other things like locating, storing, and dispensing food and other resources. The dead would begin to pile up quickly, bearing in mind that we are operating on the premise of a very rapid[17] die-off reaching 99.99 percent of the population. Before the die off reached critical conditions, barring something that kills in a single, simultaneous stroke, survivors would be consuming existing goods as they moved through the countryside and settled into refugee camps. Corpses would pile up and dead, rotting corpses, even if removed to a reasonable distance from refugee centers would attract rats and other scavengers which in turn would, as those same rats would be drawn to large stores of food, make the perpetuation of disease a constant threat for some time not to mention the pilfering of stored rations. Burning the dead would be the easiest and most efficient form of disposal, but even that might be a problem if people are dying because of radiation sickness—burning an irradiated body does not eliminate the danger of radioactivity in the remains and can also contaminate the environment where cremation takes place.

The Nuclear Option Is an Apocalyptic Suicide Pact

Just as the coal, oil, and fracking industries would leave the post-apocalyptic landscape poisoned and off-limits to human civilization for hundreds of years, the nuclear prospects for the post-apocalyptic are even bleaker. If it is a problem to find someone that can patch a coal slurry dam or operate an oil refinery imagine the difficulty in finding someone that could help deal with a nuclear facility—even one as dense as Homer Simpson would be difficult to locate, though if found, would be better than no one. Even if the nuclear facilities were all properly shut down they would still possess the ability to poison the environment for a long time. The half-life of nuclear waste, and radioactive substances, means they would potentially be a fixture of the environment for thousands of years.[18] Especially as the knowledge pertaining to such things faded with each post-apocalyptic generation. Making an allowance for the purposeful salvation of nuclear-trained persons—physicists, chemists, facility personnel, and so on—there are 61 currently operating commercial nuclear power plants in the United States alone, housing 99 reactors, spread out over 30 states and each one has the potential to become a Chernobyl or Three Mile Island.[19]

This, of course, does not include weaponized materials, like plutonium, which would not detonate on its own, but would, as the missiles housing the material corroded and fell apart, poison the areas surrounding them. Other weaponized materials, like Mustard or Sarin gases, could also find their way into the air making their locations deadly even if for just a short time. However, the radiation that would follow human survivors into the post-apocalypse, even if the apocalypse does not have a nuclear cause, would be enough to create short term deaths and lingering illness as well as long term genetic problems and potential mutations in other species. Moving forward one generation into the apocalyptic world, and allowing for just one natural disaster—an earthquake—that caused serious damage to a nuclear waste site, survivors may not even know to be on the lookout for signs and symptoms of radiation poisoning—there are more than 100 storage sites across the United States capable of potentially contaminating the more than forty states that harbor them.

These problems, even if generously considered innocuous for the most part, would make life difficult. For instance, it has been noted that wildlife around Chernobyl, while thriving, is measurably smaller. That is, the animals are not growing as large, in the case of deer and other traditionally hunted mammals, but they are also dying sooner and producing fewer offspring per pairing. So, though the wildlife has returned to Chernobyl, whatever human life followed it there in the post-apocalypse, or whatever human life attempted to live in similar areas, would have to harvest more animals to feed the populations which would only spur already struggling animal populations to extinction—again, generously assuming the radiation was not going to do that eventually to all life anyway.

The Apocalypse—A Chemical Romance

Chemicals, besides nuclear and fossil fuels, would continue to work against human survivors for at least a generation as fertilizers, pesticides, herbicides, and other industrial chemicals were soaked up and buried by soils or worked their way into streams and rivers and eventually the oceans. It is estimated that the eutrophic "dead zone" at the mouth of the Mississippi River, roughly 7,700 square miles stretching up and down the Gulf coast and out to sea, caused by the volume of just these agricultural and industrial chemicals, will take decades to sort itself out naturally. The effects of these chemicals are felt upstream and downstream and should not be overlooked as potential sites of long-term problems for life in a post-apocalyptic landscape. The Gulf dead zone is but one of many, and one of many that is continually growing, in and around the United States. In 2012, Robert Diaz, surveying numerous studies of hypoxic zones, identified 166 reports of dead zones.

> Coastal waters contain the vast majority, though some exist in inland waterways. A handful of the 166 dead zones have since bounced back through improved management of sewage and agricultural runoff, but as fertilizer use and factory farming increase, we are creating dead zones faster than nature can recover. There are more than 400 known dead zones worldwide, covering about 1 percent of the area of the continental shelves. That number is almost certainly a vast undercount, though, since large parts of Africa, South America, and Asia have yet to be adequately studied.[20]

One of the reasons these chemical caches in the waterways are so important is because they contribute to conditions on land and in the water that will alter the post-apocalyptic landscape. Invasive species of plants, like kudzu, will, left unchecked, choke out whole swaths of countryside waiting for evolution to provide a natural herbivore in the United States to start cleaning up on a scale large enough to make a difference. In the meantime, it, like other invasive plants that will feast on a cocktail of agricultural chemicals, will spread like a wildfire and, as they do, they will condemn the landscape by killing the very plants that would otherwise be hard at work filtering the toxins in the environment and dying to begin the renewal of soils.

Not All Alien Invasions Are Extraterrestrial

In other ways, invasive species, like the Asian Carp, will join forces with the chemical destruction of aquatic life and kill off a large number of native competitors leaving the underwater worlds fundamentally different. If the Asian Carp were to breach the canal connecting the Great Lakes to the Mississippi River, it would spell doom for a substantial number of fish species across the Great Lakes Region. Nutria, left unchecked by disgruntled humans, would work tirelessly at the banks of streams and rivers spreading northward altering the courses of waterways along their path of migration. These are but a few of the species that would contribute to the alteration of the post-apocalyptic landscape, but their success would be tethered to the conditions we have already created for them to take advantage of in the event of an apocalypse. This is especially true where some of these species, like their human counterparts, may be able to outlast even the direst of apocalyptic conditions.

Of course, alongside invasive species would be a whole cadre of newly feral dogs and cats and exotic pets escaped or loosed upon the world by sympathetic owners just before they died. These animals—not to mention what might come of an environmentally zealous group of persons opening the cages at zoos and circuses—would present new predators to the landscape that would also be fleeing the toxic conditions we would have created. So, as humans drew closer and closer together into more hospitable areas they would be followed by a host of dangerous predators which, before long, as ammunition becomes more scarce, we would no longer be able to defend against to say nothing of hunt.

The Sum of All Fears

All these problems have found expression in the dystopias of popular culture. However, where that has been the case each, to greater or lesser degrees, has been treated as the primary or penultimate threat which, once overcome, leaves humanity free to rebuild in relative safety. However, taken together it becomes clear that life in the United States would become nearly impossible faced with an apocalyptic death toll. Worse, these are only the vanguard of problems that would make life untenable. There are myriad smaller problems which would, in their own way, become equally disastrous for the survivors compounding the problems of the apocalyptic landscape even further. For instance, plastics and other refuse would continue to choke the seas and pollute the land. Working their way up the food chain it would be a long time before aquatic cuisine could be consumed without traces of plastics and heavy metals. This is, of course, if the sheer amount of debris in the oceans does not manage to choke oceanic life to death when combined with desalinization from glacial melt and warming waters. Algae blooms would work in tandem with plastics, a waste product we have yet to determine the rate of biodegradability for and which would continue to move through the landscape and waterways via turbulent storms and erosion, especially where landfills were exposed to such things.

All of this would be happening in the context of a world built by the lowest bidder with the cheapest materials. As the infrastructure gave way from lack of care, battered by storms, and destabilized environmental conditions, the world would become a place of decay and rot and industrial waste to rival any dystopian post-apocalyptic vision yet. The environment ought to factor into these narratives in a more central way, especially as a functional part of the social critique, if such dystopianism is ever to motivate changes to the way we are currently living. There are plenty of people who think they would be able to live, not just survive, in a post-apocalyptic world primarily as a result of misleading pop culture renderings of the post-apocalyptic environment. The concern with various futurisms that take as their setting the post-apocalyptic is that they have been presented as, and have been accepted as, desirable *precisely* because they offer the clearest and quickest path to a potential utopian understanding of the world.

Though it seems reasonable to assert that a future society might have an apocalyptic pandemic narrative built into its socio-cultural history our present dystopian literature betokens other likely candidates as well. The major problem with using dystopian literature to explore the possible apocalyptic is that most dystopian literature does not require a full breakdown of society to achieve its ugliest conclusions even when the dystopia is the product of a disaster. Disasters, especially large-scale disasters, bring into intense focus matters of human adaptation and disasters can reveal basic aspects of how a society conforms to the features of its physical environment including the crux of its survivability.[21] Though, again, it must be remembered that large scale disasters are not the equivalent of apocalyptic conditions.

3

Pathogenic Shaped Futures, Part I

Annihilation and *The Walking Dead*

You Step Outside, You Risk Your Life

In Robert Kirkman's *The Walking Dead*[1] we are treated to an epic narrative of life after a super-virus decimates humankind. The main protagonists are located, thus far in the telling of the story, in a triangular area where Cynthiana, Kentucky, the metropolitan area of Atlanta, Georgia, and the suburbs of Washington, D.C., anchor the points. In the story, Kirkman never fully divulges what causes the outbreak,[2] but we get an idea of how quickly it spreads throughout human society and, obviously, it was thoroughly devastating to the population achieving something very near the earlier discussed 99.99 percent die off rate. Of little interest to my analysis is the reality, or lack thereof, of a zombie apocalypse. Regardless of whether zombies could exist the story arc is still relevant because even given just a mass die off social norms would be broken and life without zombies would still be just as difficult because the same social and environmental challenges would exist. So, while an analysis of *The Walking Dead* will require occasionally discussing zombies, for the most part they are irrelevant to how such a decimation of human society would affect the environment.

At the outset of the story there is no hint of the looming apocalypse though one can easily imagine the first cases of unusual deaths were being reported around the country as Rick Grimes was being gunned down. The first page of the story is the only time that Kirkman allows his audience to see the world as it was before the zombie apocalypse. Rick and Shane, his partner, are engaged in a gunfight with a criminal hell-bent on not going back to jail. The gunfight takes place on the outskirts of Cynthiana, on some country road, about ten miles from town. The scene is classic American pastoral: small buildings silhouetted in the background barely breakup the horizon, horses frolicking in a nearby field hemmed in by a wood slat fence, a late-summer sun illuminates the landscape, and, despite the black, white, and grayscale rendering of the novel, one can easily imagine the muted pastels of a late summer sunset in a variety of fruity colors.[3] It is here, in this rather mundane rendering of the world, that Rick Grimes is shot and the world as he knew and experienced it would cease to exist.

If Rick were able to recover mostly or, best case, fully, from his gunshot wound to awaken to a world already turned into a zombie nightmare it is a fair estimate that only a few weeks have passed. The safe assumption being roughly six weeks given the location of the gunshot wound—the right side of his chest—and the timetables for a typical recovery, operating on the assumption that the wound was serious but superficial, allowing that it did not damage any major organs or arteries that could not have been fully repaired. His recovery time does not account for Rick's coma which could have lasted longer than the time necessary to recover from a gunshot and would allow for the extension of time between his being wounded and waking up stranded in a hospital. Obviously, Rick awakens weakened, but able to move around, dress himself, and, ultimately, fight off the zombies trapped in the hospital with him. So, it is fair to speculate that he has fully recovered from the gunshot wound and his coma lasted a bit longer than necessary for that to happen, say two weeks longer than his recovery from the gunshot alone would have taken.

The timeline between Rick being shot and waking up abandoned in the hospital is important because we know that is the amount of time it took for the apocalypse to get bad enough that people were evacuating their homes and towns in search of safer places. When Rick finally catches up to his wife, son, and partner outside Atlanta, Lori, his wife, tells him that it appears he was left in the hospital to fend for himself less than, or at most, a week after she and Shane took off for Atlanta.[4] So, Rick is shot in late Summer, recovers in the hospital for roughly six weeks, and then lingers on in a coma for maybe a week or two before waking up—sometime in October. In that amount of time the zombie apocalypse is in full swing and the world should be undergoing the changes that would accompany a massive and rapid population decrease and mass evacuations.

When Rick leaves the hospital, the world is already showing signs of disuse and disrepair that one might expect following mass evacuations amid accelerated and near total population destruction. As Rick exits the hospital we get our first glimpse of an apocalyptic world: we know it is autumn because the trees have no leaves, there is litter and various detritus accumulating in and alongside the road, the grass has obviously not been mown in some time, but all in all the world does not look much different than any typical, that is, non-apocalyptic autumn day. Rick makes his way across town and as he nears his house we get a different glimpse of the apocalyptic conditions of an evacuated town.

As Rick approaches his house the scene is more dire than the area around the hospital would lead one to believe. There are burned out houses, obvious signs of looting, broken windows, destroyed property, garbage piled all around, homes ransacked, and, again, plenty of unmown grass.[5] There is a clear sense of disorder in the evacuation of Cynthiana and one can only imagine how, in that disorder, precautions were not taken to secure potential hazards. Obviously, this lack of concern stems from two sources; on the one hand, lots of people were dying very quickly and resurrecting leaving survivors little choice but to flee and, on the other hand, in taking flight people were clearly concerned about their immediate well-being and not their long-term well-being. The first is made clear through the depiction of what remains of Cynthiana. The second is made clear through the discussions Rick has with the survivors around Atlanta who continue to believe that some sort of rescue will eventually take place and that the government is doing what they can to bring order back to society.

In the pandemonium that would surely ensue as people fled for their lives both running from zombies and searching for safe havens how many oil refineries were properly shut down? How many chemical factories were shut up tight? The chaos alone, including looting, which Kirkman makes clear was taking place, especially in the early stages of the pandemic, would have left the world on the precipice of massive and near total collapse. There is a reasonable question of how long it would take for the world to fall apart in tragically uncontrollable ways, but Kirkman has already given us a snapshot how people in *The Walking Dead* universe would respond to apocalyptic conditions. Houses burnt, broken, and destroyed for the goods they contain make it a near certainty that business and other buildings faired at least as poorly, so that evacuated places and the evacuation routes are sites of destruction, and, perhaps, more aptly conceived as self-destruction.

This is confirmed by Kirkman in his representation of Cynthiana, which we are shown is laid to waste as Rick leaves for Atlanta. Wrecked cars, broken windows, and debris of all types litters the streets as he navigates his way out of town. On his way to Atlanta the interstate is littered with abandoned vehicles and debris. Rick stops at a gas station in Georgia, north of Atlanta, and it is shown to be ransacked as well. Atlanta, upon Rick's arrival, is a wasteland just like Cynthiana. This is to be expected as Atlanta was the location that early survivors were flocking to for protection. It stands to reason that once tens of thousands of survivors were added to the metropolitan areas' millions of residents the situation in the safe zone had to have spiraled out of control rather quickly to say nothing of violently. Not just in terms of the zombie threat that was lurking amongst the survivors, but in terms of scarce resources for sustenance, defense, survival, and self-care (medications, clothes, shelter, etc.).[6]

Atlanta is not just looted like Cynthiana, it is desolate and destroyed, in addition to the usual evidence of looting and the expected accumulation of garbage in the streets, we see graffiti and evidence of a higher level of chaos is imagined having taken place. This is reinforced with the presence of tanks and National Guard troops once attempting to defend survivors and, in the end, destroy the zombies in self-defense. Glenn, after he rescues Rick, reveals that Atlanta fell in a week and, so, given the depictions of the two cities thus far one ought to conclude that the difference between small towns like Cynthiana and bigger cities like Atlanta is the difference between run-of-the-mill looting and abandonment, on the one hand, and warzones, on the other.

Except that from the point Rick arrives in Atlanta, to the moment he is saved by Glenn, including their flight across the rooftops to reach the outskirts of the city, nothing we are shown is burning or has burned. That is not to say that fires did not occur, but the expectation is that between the looting, fighting, dying, and general disarray that must necessarily prevail in such circumstances, there is every reason to believe that Atlanta would be burning, or already have burnt, to the ground on a scale that would rival General Sherman's handiwork. Bear in mind that Cynthiana had burned out buildings in both residential and commercial areas, so it stands to reason that Atlanta would be worse off especially considering how quickly Atlanta fell. Yet, every time the survivors venture into Atlanta it is represented as a veritable enter-at-your-own-risk cornucopia that has managed to largely escape the tribulations suffered by smaller cities.

Once outside the imminent danger of the city, Glenn and Rick catch their breath in yet another serene pastoral environment. Never mind that the countryside is not afire as the city ought to be, but more than that the wooded area is *clean*. There is no litter or debris built up along the base of trees, the grass is the perfect height for an unmanaged autumnal meadow, the trees are also denuded of leaves, but at no point between exiting the city and Rick's arrival at the survivor's camp is there any indication that an apocalypse is underway. In fact, life in the survivor's camp is pleasant for the most part with the obvious exception of living in the open amongst zombies. They have plenty of canned goods, a shower, and, thanks to Glenn's resupply runs into Atlanta, the survivors even have some luxuries such as toilet paper, laundry detergent, and candy bars.[7] All of which is perfectly believable at this stage in the story except for the fact the group has not made an effort to collect guns, ammunition, or any adequate supplies for their defense.

The day after Rick's arrival he and Shane go hunting while the women stay behind to wash the clothes. Though one of the women in the group, Donna, makes her dissatisfaction about the division of labor clear, the other women, Lori included, seem to think that such things are to be expected. Three things stand out about this scene early in the apocalypse. First, the survivors have generated a social context that mirrors life before the apocalypse with leadership and a clear division of labor within the camp. Further, while there is concern for how such a reversion to a more primal way of doing things could affect the return to civilized co-existence, the social order seems to remain intact in spite of the large-scale breakdown of society. This is made explicit, first, when Shane tells Rick the situation, such as it is, "won't last long" although it has been going on for at least a month or more and the situation has gotten pretty dreadful.[8] Secondly, this sentiment is reinforced by the women's conversation about whether or not the division of labor in the survivor's camp is about "women's rights" or "being realistic and doing what needs to be done."[9] Both conversations harken to an optimism about the eventual return to the way things were and what is to be expected in the meantime.

The third important take away from this scene is how accommodating the environment is with respect to the needs of the survivors. Rick and Shane go off into the wooded area surrounding the camp, and separating them from Atlanta, and the area is apparently untouched by the effects of the apocalypse. Bear in mind that the survivors camp must be close enough to the city that Glenn can make it to Atlanta, infiltrate the city, search for supplies, exit the city, and return to the camp *all in one day and on foot*. So, there is an expectation that the area around a city as destroyed as Atlanta would show some signs of cross contamination. Moreover, the woods have a stream within earshot of the camp, an easy walk away, where the women go to wash the clothes. There are two separate depictions of the creek and neither one shows any sign whatsoever of the ill effects of the fall of Atlanta.

Never mind that the stream bears no signs of even minimal pre-apocalyptic contamination at all from its proximity to so large a metropolitan area. One could almost forgive this if the women were going to launder their clothes at the pond nearby—which we know exists because Shane tells Rick the shower is pond water. Presumably, the pond has been thus far spared contamination, but the stream is a flowing body of water and regardless of whether it is flowing into or out of Atlanta, given the condition of small towns, the routes of evacuations, and Atlanta proper, there is every reason to expect that the stream would be noticeably polluted at best and unusable at worst.[10]

The world has ground to a halt in the aftermath of rapid onset zombification of the population and yet, here are a bunch of survivors attentive for errant zombies that may do them harm, but who are otherwise, quite literally, camping in the most pre-apocalyptic sense of the word. The civilized parts of the world may be crumbling—institutions, infrastructure, social mores, etc.—but the environmental aspects of the world have merely stagnated, or improved in some ways, making it possible for a reasonably comfortable semi-civilized lifestyle to continue. Rick and his hardy band of survivors, though confronted with the regular need to go collect canned goods from homes and stores, to hunt as best they can, and be ever watchful lest a zombie bite them, are able to move about the world quite unhindered by the collapse we know ought to be taking place rather quickly where they are located.

Words Can Be Meager Things

Shortly after Rick's arrival the survivors are forced to move their camp. At this point in the story it is visibly well into the winter season. However, the representation of the seasonal change seems to be more a device to mark time than an actual representation of anything climate related. *The Walking Dead*'s timeline at this point in the story is late summer to early winter and this has been casually marked out by the change in the setting from a shooting on a sunny summer day through naked trees and long sleeves to jackets and falling snow. The falling snow is not terribly out of place in Northern Georgia and there is every reason to believe the survivors are in that vicinity. Rick traveled south to Atlanta along the I-75 corridor from Kentucky. He meets up with Glenn shortly after wandering into the city and he and Glenn safely make it to the camp on foot. Also, given how underprepared the survivors are for staging a long-term post-civilization existence, when they do move camp they are, reasonably, moving south.

Winter in Georgia does not produce a lot of snow—even in the northern part of the state—and in the story, there is no real accumulation of snow on the ground, so, flurries most likely are what betoken winter's arrival. Yet, following the survivor's sojourn through the apocalyptic countryside, presumably heading south, *the weather worsens*. Heading out of the camp Rick tells Dale "thank god the snow has let up" and in the bird's-eye visuals that accompany their flight from the camp there is a considerable amount of snow on the ground.[11] Enough that driving is difficult and it is easy to get the impression that road conditions would have been treacherous even without the lingering effects of apocalyptic roadways—deserted cars, debris in the road, and *frozen* zombies.

Perhaps it does not have to be terribly cold for a walking dead person to freeze, and not all the zombies are frozen, though most of them are slowed by the weather, but the fact remains that the survivors are facing weather conditions that are completely out of place in central and southern Georgia. This, however, is not a fact about the weather in the novel, if anything, it is merely, as I stated above, a device to mark the passage of time, and most people expect winter to be wintery. It helps keep the story interesting that the weather conditions make life a little harder for our intrepid survivors, but that misses

3. Pathogenic Shaped Futures, Part I

the point entirely of what could be happening to the climate six months into a population reduction like the one represented in *The Walking Dead*.

In fact, as they search the Georgian countryside for a suitable place to establish a new base to operate out of two things are made quite clear: this is a pretty harsh winter storm and resources are scarce across a countryside that has no corner unaffected by the recent apocalyptic events. The group still finds many houses burned out, heavily looted, or otherwise untenable for a group of their size—not to mention that they picked up three additional people during their haphazard wonderings. When the group finally finds the well-stocked, gated community of Wiltshire Estates, the storm has broken and, once again, there is a glimpse of the apocalyptic world as tolerable, bountiful, and livable, again, with the exception of the hazards presented by the ever present zombie threat.[12] The fact that several people in the group are from Georgia, but none of them feel compelled to comment on the strange severity of the weather—a blizzard has not occurred in Georgia since 1993—further reinforces the use of the weather to mark the passage of time.

The weather in Georgia is not severe enough to warrant the level of hardship the snowstorm in the novel represents. The weather in Georgia has the ability to turn rather nasty after reasonably high winter temperatures. In January 2014 Atlanta had temperatures in the 60s for a week before a winter storm hit grinding city life to a halt for two days. That was the now famous "Snow Jam" of 2014 and it delivered a walloping 2.6 inches of snow and ranks as the 20th worst storm in Georgia's history. The blizzard of 1993 dropped a much more impressive 4.2 inches of snow. Never has the Atlanta area had a snow storm produce 10 or more inches of snow the closest being 8.3 inches in 1940.[13] So, obviously, a big snow storm is possible, but the likelihood is so slim that the fact the Georgia natives make no comment about the extremity of the storm leads to the simple conclusion that the storm exists because snow is the hallmark of the winter season and it is an easy way to complicate the lives of the survivors.

This matters primarily because, in a post-apocalyptic world, after six or more months of no human oversight of dangerous industries that provide the fuel for the engine of global climate shift, it is quite likely that weather patterns and climates could become worse rather quickly. So, either this snowstorm is a Georgia anomaly, or it represents the beginning of something far more difficult for the survivors to face as time continues to pass. Winter continues raging on around the survivors and becomes especially problematic after their hasty withdrawal from Wiltshire Estates. Wiltshire Estates was well stocked with a variety of provisions the group needed, including food, but the survivors are forced to leave it all behind. As a result, the abnormal winter conditions make finding food difficult driving the group to look for scraps in increasingly dangerous and useless locations.

Tyrese remarks on the slim pickings and general state of destruction they encounter as they continue to look for a permanent shelter. He says to Rick and Glenn, as they survey the empty contents of a destroyed general store, "anything that wasn't taken was destroyed. Why the hell would anyone do this?"[14] To which Rick responds that there will be a lot of hunting in their future. So, again, the group, which was hunting on a limited basis outside Atlanta, is now scouring the countryside for food on the hoof. Once again, members of the group are shown going on excursions into wooded areas that are seemingly safeguarded against the type of wonton destruction that developed areas have been subjected to wherever the group travels. It is on one such occasion that Rick's bedraggled band of survivors is brought into violent contact with another group of survivors as Carl, Rick's son, is accidentally shot by the hapless Otis who also happens to be out hunting for food. This unhappy collision of surviving groups brings the Atlanta group to the farmstead of Hershel Greene.

The World That We Knew Is Dead and This New World Is Ugly

The first look at Herschel's farmstead is reminiscent of the pastoral scene of Rick's shooting. Herschel has a home and outbuildings in good working order, livestock meandering in a nearby field, fields lying fallow in anticipation of spring's arrival, and, the only noticeable difference between the two pastoral scenes, a functional, if somewhat flimsy, protective fence surrounding his property to stop zombies from getting too close.[15] Overlooking the good fortune that befalls the Grimes family after their son is shot—Hershel is a veterinarian capable of saving Carl's life—the Atlanta survivors are equally lucky to have stumbled upon Hershel's family. Shortly after stabilizing Carl, the two groups sit down to eat dinner together. Not only does Hershel possess the space to seat twenty-three people, but they are fed reasonably well even with "the markets being closed."[16] Hershel demonstrates that he is both a successful farmer and a competent manager of resources. One is left wondering why Otis was out hunting with so much food readily available.

Every scene at the farm is presented as an American idyll. The farm is fully functional housing pigs, cows, chickens, and horses, not to mention the eight survivors in Hershel's group. Life at the farmstead continues as if nothing terrifying and deadly were besieging the place. The zombies are as inconsequential to Hershel's existence as the markets being closed is a nuisance. At this point in the story it is midwinter, roughly a little more than six months since Rick was shot, the survivors are somewhere in southern Georgia, and though human existence is seriously imperiled, life continues as if a massive population decimation were but a bump in the road. This is the kind of human-centric anti-environmental bravado that establishes a parallel narrative in actual public discourse about the ability of humans to survive such an occurrence. Not just to survive it, but to overcome it, made clear by the fact that Hershel has been keeping an eye on his neighbor's property even as he has them locked away in his barn for safekeeping until they manage to recover from death.

The ability of Hershel and Rick to maintain an above average level of existence for sizable groups in Georgia is quite outstanding. Consider that the Atlanta metropolitan area houses just shy of six million people. Atlanta is one of the largest metropolitan areas in North America, coming in ninth on the top ten list in the United States, just behind Miami. If Atlanta was chosen as an evacuation destination primarily because of its size and the resources available in and around the area it is safe to surmise that, in the days and weeks leading up to the point where the zombie outbreak overtook the survivors, the population of the metro area could have easily doubled. Obviously, a large metropolitan area like Miami would also, potentially, have been designated as an area for people to go to as they evacuated. The American southeast (defined as South Carolina, Georgia, Florida, Alabama, and Mississippi) is home to roughly ninety million people *before* people start pouring into the Atlanta area from outside that region. So, granting a doubling of the Atlanta metropolitan population seems fair and a fairly substantial underestimation.

Given a population of approximately twelve million, and granting that ninety percent of those people are dead, or walking dead sequestered within the confines of Atlanta's city limits, there are 1.2 million zombies on the loose in the greater Atlanta metropolitan area. Given that the action of *The Walking Dead* takes place within roughly fifty miles of Atlanta's city center, there should be somewhere around 480 walkers per square mile. Ignoring for now that Rick's hardy band of survivors lived within walking distance of Atlanta, yet encountered very few zombies, how is it that Hershel's farm, bristling with the activity of ten people and lots of livestock, has managed to avoid a substantial number of zombies? There is an argument to be made here that the environment is "cleaning up" the zombies, trapping them in natural obstacles, funneling them into urban centers, or eliminating them altogether. But, if that is the case, why is the environment, as depicted, so healthy and devoid of the dead?

After being evicted from Hershel's farm, Rick, and his fellow travelers, once again take up the back roads of Georgia and, once again, there is nothing to indicate that the environment is suffering an apocalypse. While their journey seems to be an aimless rambling, it is during this time that the group discovers a salvific prison nestled in the foothills of the Georgia backcountry, largely isolated from nearby towns and residential areas. This circumstance is signaled first when Rick announces to the group they are "out of gas" and since they "haven't seen any stranded cars in a while" they are going to have to go scavenging whether they like it or not.[17] Moreover, Rick instructs everyone to "spread out, look for cars, anywhere," adding "if you see any nearby houses, let us know, there's got to be something around."[18] That the group is some distance from developed areas is obvious from the discussion and the depictions of the landscape they are setting out to search.

You Ever Heard About the Broken Window Theory?

The snow has clearly melted away, though the group members continue to wear winter style clothing, signaling, again, a change in season from winter to spring. The forest seems a little leafier and, when the group get their first good look at the prison, Kirkman provides a large panoramic hilltop view of the structure and the surrounding area. Wide ranging fields in all directions are presented as manicured despite having gone several months without upkeep. The prison, and especially its fences, are in incredibly good shape, and the only genuine concern facing the group as they look in bewilderment upon their future home, is the abundance of zombies roaming about primarily *inside* the prison's fences. Nothing in the depiction of the prison landscape suggests apocalypse, or a hard winter, in fact, the whole scene appears to have avoided the apocalypse altogether, except for a few pesky zombies that serve as a reminder. Upon closer inspection, there are some scraps of paper littering the area outside the prison fence and, though the prison fence is slightly broken in a few places, everything else seems to be in fine working order, leading Carol to speculate that they "could rebuild—make a *new life*."[19]

Once the group sets up in the prison compound the idea of a new life begins to take deep root as the focus shifts from survival to rebuilding a social environment. Conversations about leadership structure, the death penalty, relationships, and, eventually, defense of the prison, slowly become the central narrative.[20] The majority of the action takes place indoors or within the prison grounds. Hershel and his family are relocated to the prison and, as a result, the prison yard is converted into farmland. The additional people make it possible to transform other aspects of the prison into functional contributions to the social life of the survivors. Only once after the group arrives at the prison does the environment around the prison play the major role of the setting.

Shortly after a tragically unnecessary suicide scene Rick realizes that all dead people come back, not just the ones that are bitten. To verify his theory, he takes a motorcycle from the prison and drives to Shane's grave near their original campsite. Lori laments, shortly after his departure, that he must be going further than Hershel's farm which is a mere four hours away by automobile. Rick is gone for a full day, perhaps a little longer, but the important aspect of his journey is that he traverses all the places they have already been through so there is a little retrospective given on the whole saga up until this point. What can be seen is that literally nothing has changed. The grass is ankle high to Rick when he is shown dismounting the motorcycle. There are no signs of fires either in his immediate area or black columns of smoke in the background. No destruction of the areas the group traveled through has occurred since they vacated the areas.[21]

If the ever-present zombie pests were erased from the storyboards there would be nothing to indicate that Rick was traveling across an apocalyptic landscape. And, given that the group is now trying to rebuild, redefine, and restart civilization, there is a deep-seated understanding at this point that the only environmental factor working against the survivors is the zombie threat. The survivors have food, shelter, safety, numbers, plenty of resources to sustain and, in some cases, improve their lives. There is never concern about the state of the world beyond an ever-present recognition of the need to deal with the dead. This story arc holds even after the survivors in the prison encounter the group of survivors in the Woodbury community. But, it is not until they do encounter the Woodbury group that anyone goes out into the world beyond the confines of the prison—well, no one that the story follows, because Andrew is banished, but beyond a shot of him fleeing the prison, there is nothing detailing what he experiences out in the open. Similarly, Michonne's arrival at the prison includes nothing about where she has been or what she has seen, and still, nearly a year into the ordeal the prison's surrounding countryside continues to be a beautiful and nonthreatening vista.

I See a Chance to Make a New Start

The next large-scale depiction of the environment is provided in the scenes between the moment the survivors spot a helicopter overhead and the moment they come into contact with Woodbury. Rick, Glenn, and Michonne leave the prison in search of the downed helicopter in hopes of finding people with knowledge of a rescue or at least some information on what is going on in the world. They leave the prison in a car, and travel down roadways that are not choked with debris or any obstacles that might impede their speedy trip to the crash site. There are no downed power poles, though the power lines are starting to sag a bit, presumably after the electricity stopped coursing through them. As Dexter pointed out to Rick when they first found the prison, the electricity had been out for several months.

The grass in the meadows along the roadside is still barely taller than ankle height, and even this far out from the prison, there is nothing to suggest apocalypse. It is easy to think that maybe the growing horde of zombies encircling the prison are tramping down the grass as a way of explaining the short grass around the prison. However, the zombies this far away from the prison are much more aimless, more spread out, and far fewer in number to account for the lack of unchecked growth the areas Rick, Glenn, and Michonne are traversing. Even as the trio is walking through the area in search of the Woodbury survivors it is easy to get the impression that an afternoon with a weedeater could put the landscape back in proper order.

Most of the action at Woodbury is indoor as well, but when the Governor finally heads to his apartment after a night of terrorizing Rick and his companions, we get a look at the town stronghold as the sun comes up. Woodbury is four square blocks of protected, orderly, clean, safe space. There are plenty of people out and about early, kids at play, and a general atmosphere that nothing is wrong with the world besides the zombie threat safely blockaded outside the makeshift walls. Of course, after the imprisoned trio escape they cross over the same terrain that they covered going to Woodbury and, as expected, they encounter nothing of note that would indicate the world was experiencing an apocalypse. The wide shot of the world that Kirkman provides when Rick chases down Martinez to prevent his return to Woodbury is another picturesque view of the American pastoral that has become the common depiction of the world throughout the early telling of the story.

Throughout the remainder of the first compendium the survivors continue as though they are well on their way to returning to civilization. Rick, sitting indoors with Lori, comments on the heat as spring rolls over to summer which, again, just like the snow, is used to demarcate the change of season and not to address the sweltering heat that the people must be suffering in the deep south, in the heart of summer, in an environment that was substantially altered prior to the apocalypse.[22] Developments and urbanization have changed the heat dynamics of industrialized societies such that a mere year would not be enough to mitigate the exacerbation of the extremes of summer temperatures. Roughly 50 miles out of Atlanta, the survivors would be experiencing the summer of South-central Georgia, and the average high temperature would hover around 90 °F and that would be intensified by dwelling in a concrete bunker without air conditioning. Axel, laboring in the large gardens Hershel has planted inside the prison fences also complains about the heat, but like Rick's mention, it is to move the story along its timeline more than to depict the harshness of life in the deep south without the conveniences of modern technology—it is after Axel's complaint that Hershel indicates that it must be July, maybe August, and in doing so alerts the reader that the story has covered its first calendar year.

More miraculous than anything else thus far in the story used to indicate

3. Pathogenic Shaped Futures, Part I   61

that the survivors are well on their way to ushering civilized society back into being is the moment they manage to construct an artificial leg for Dale to use to get him off crutches.[23] The story settles into a narrative that focuses so exclusively on the lives of the survivors as they attempt to return to civilized society that there is no depiction of the environment worth discussing until the Governor shows up outside the gates of the prison to wage war on Rick and his group. Even then the focal point of the storytelling is narrowly fixated on individuals, individual actions, and defense of the prison from inside the prison. The only good views of the environment are given at the Governor's staging ground for his attacks on the prison. But these areas are so close to the prison that they not only offer little new to the understanding of how the environment might have been affected by the apocalypse, they actually reinforce the already pervasive idea that the environment is either statically stuck in whatever state it was in on the day the apocalypse started or it is improving on its own.

The obvious response to such an assertion is to take a more generous approach to the way that Kirkman and his creative team have gone about structuring the story. After all, the events of *The Walking Dead* up to this point have taken place in the span of one year in a very limited area. So, it stands to reason that the world has not actually started to breakdown or has but the effects of the breakdown have not managed to make their way downstream to the somewhat isolated locales that the survivors inhabit. This is definitely a generous view given the analysis of the problems stemming from the original campsite location near Atlanta. However, granting that one year is just not enough time for things to get truly awful, or that the conditions have been favorable preventing the worst of the anticipated ill effects, or that the survivors have thus far been located in areas where the worst outcomes of a massive population reduction just have not been visible or widespread enough to be representative, these perspectives account for the missing environmental fallout of a near total population decimation. Hence, in the second compendium things should be an order of magnitude different since the individuals that survived the Governor's raid are now back out in the open once again traversing the Georgia countryside and, eventually, making a long cross-country trek to Washington, D.C.

This Is Our Extinction Event

Immediately following the loss of the prison sanctuary Rick and Carl pass, on foot, through wooded areas and residential neighborhoods. The scene in these areas is unchanged from what they encountered prior to settling into the prison. When the twosome finally take shelter in an abandoned house, due to Rick's ailments, the house looks, externally, as though it is ready to go on the market.[24] Far from having been ravaged like Rick's own house had been in Cynthiana, this home is in good working order. The subdivision the house is in looks roughly equivalent to how other residential areas looked with debris and wrecked cars piled haphazardly in the streets, but for all intents and purposes the whole neighborhood appears exactly the way it must have when it was first evacuated months, if not more than a full year, prior to Rick and Carl sheltering in it.

While they are holed up in the house recuperating and attempting to determine what to do they venture out into the neighborhood and into the nearby woods to scavenge and hunt for food. At no time during this sequence of events is there any indication that things have gotten worse or could get worse with the exception of the immediate concerns one would expect to be vexed by in such a situation—lack of food, water, ammunition, etc. They do, on one outing, witness a man commit suicide by zombie and they discover his trove of goods, which they keep, but for the most part they are inhabiting a world that is not merely resilient, but remarkably unthreatening.

When they finally vacate the premises in search of something safer and, presumably, an area with a greater amount of scavengable goods, Rick lets Carl drive the car. There is obviously some comedic relief in the episode since Carl is both too little and too young to operate a vehicle, but it is after he nearly crashes that Rick explains his motives for making Carl drive the car. Rick tells Carl that "it's ... important ... that [he] learn things like this" in the event they get separated. Carl sees the red herring and points out that he understands that Rick thinks he might die. Rick acknowledges this fact and says he wants to be sure that Carl can take care of himself. So, a year into the apocalypse, when most cars are out of gas and gas is a hard to come by commodity, Rick wants to teach Carl how to drive as if this is a skill that a nine year old boy needs to know in order to survive the apocalypse should he ever find himself on his own. Even if he could scrounge up the gas on a regular basis, he could not repair a vehicle, or, like Glenn, hotwire one in a pinch. Why is this skill set at the top of Rick's list of important things to teach his son? Even if it is taken as a given that this skill is important, how long will gas make it viable? Kirkman has already shown how difficult it is for adults to find gas in the apocalypse, so clinging to this outdated notion of necessity betokens a mindset still fixed on the idea of a return to civilization.

This is juxtaposed nicely against the reunion of Rick and Carl with the other survivors of the prison fiasco when Glenn and Maggie, astride horses, stumble upon Rick. The need to procure and maintain transportation in the apocalyptic landscape must be putting pressure on any survivors at this point. Transportation is the primary concern of the new characters introduced into the story—Abraham, Rosita, and Eugene—as they argue against fixed camps or lingering too long in any one place. Yet, again, here is a group of survivors moving around in a gas guzzling, two and half ton truck, with an average fuel economy of 10 miles per gallon.[25] Not only do the new group of survivors drive a vehicle that is a drain on their resources, once the two groups band together to make the trek to Washington, D.C., they take an additional vehicle with them, and, eventually pick up a third. Along the way gasoline is the primary concern of the group while food and other resources are readily picked up along the way at abandoned towns that dot the map along the route they are following, and, additionally, they have a concern for finding a vehicle more suitable to sleep in as the weather changes, yet again, from summer to fall.

End of the World Don't Mean Shit When You Got a Tank

During the trip from Hershel's farmstead eastbound to Washington, D.C., the world that the survivors travel through never changes in its makeup or representation. The cars along the way are wrecked, buildings have the appearance of having been looted, broken windows and doors on busted hinges are the primary markers for such activity, and there are zombies all around in various stages of decay. Yet, the grass is still ankle high, the landscape sans zombies is remarkably pristine, though to be fair there are tufts of grass and weeds growing in the cracks and crevices of the roadway, but the roads remain completely passable beyond the obstacles created by long abandoned cars. The group is never shown having to maneuver around an obstacle to their path more frustrating than the occasional car or car pileup. In fact, as Rick's newly constituted group travels to Washington, D.C., right through the heart of Appalachia, they never once cross paths with anything more threatening from the environment than an abandoned town, zombies, or other lecherous survivors out to make a killing in a world with no laws.

Leaving Hershel's farmstead, the group travels north and skirts Atlanta to the west. When they finally stop to pull out a map and plan their route of travel they are at an intersection with Interstate 75—the same road that Rick took into Atlanta—and Eugene says they have, "gone far enough north to miss Atlanta," but he insists the group needs to head east and avoid traveling on the interstate.[26] Going east will eventually allow them to pick up Interstate 95 and follow it north right into the heart of Washington, D.C. Given these clues, and the fact that Rick believes they are roughly 250 miles or so away from Cynthiana, it is a safe bet that the group has stopped somewhere near Cartersville, GA. There is a route from Cartersville to Florence, S.C., where their path would lead them directly to Interstate 95, avoiding major highways at the same time—which is what Eugene wants to do.

Before plodding their way across eastern Georgia and South Carolina, Rick and Abraham detour back to Cynthiana to, once again, raid the police station for supplies. Cynthiana looks remarkably as it did when Rick first pulled out to head to Atlanta. It is, perhaps, a bit grungier, and Dwayne has certainly suffered the effects of an apocalyptic year in Kentucky, but besides the additional trash blowing around in the streets, and the increase in zombie activity, there is not a whole lot of change to the environment. This is the only substantial scene in the story where a before and after look is given to the reader and there is very little to indicate that anything has really been destroyed by the lack of human oversight beyond the expected state of disrepair that follows from months of neglect.

Traveling north to Cynthiana and, again, when traveling south to rejoin the group the roadways are usable, there are no environmental hazards plaguing the group—either group—during this period, and no background indicators to suggest that awful things are happening around these people as the world *constructed by humans* crumbles. On their return trip Rick, Abraham, Dwayne, and Carl encounter a herd of zombies. In the frames of the graphic novel that depict this scene Kirkman has drawn a panoramic vista of the I-75 corridor as seen from the crest of a hill. If the zombies were erased from this frame it would look like a garbage truck drove down the interstate leaving a trail of trash behind it. As the group flees from the zombie herd they run across fields, through a little wooded area, and alongside the roadway, areas that have not been cared for in more than a year, the grass is ankle high, there is no evidence of wild fires, nothing that would suggest that 99 percent or more of the human population rapidly died off.

From this point in the story onwards the group is outside, traveling through fields and wooded areas paralleling the roadways they are following and there is literally no change to the environment despite crossing through much of the southeastern United States. From the farmhouse and temporary gas station headquarters where the majority of the group was gathered waiting for Rick and Abraham to return from Cynthiana, all the way through hiding out in Father Gabriel's church, it is impossible to decipher whether the apocalypse has been going on for more than a year or just started, so unchanged is the world the survivors are living in. After the events at the church the story fast forwards to somewhere just south of the Maryland border.

Although the group of survivors is struggling to find enough to eat that is quite clearly their only concern beyond security against the ever-present zombie threat. They have, as a fairly large group, crossed a significant portion of the southern United States and never encountered the fallout from the defunct coal industry despite having crossed the paths of roughly 24 hazardous coal sites on the journeys the group has undertaken.[27] According to Earthjustice, "Coal-fired power plants generate about 140 million tons of fly ash, scrubber sludge, and other combustion wastes every year. These wastes contain some of the earth's most deadly pollutants, including toxic metals that can cause cancer and neurological harm in humans [and] coal combustion waste sites are known to have contaminated groundwater, wetlands, creeks, or rivers."[28] Of course, it is important to bear in mind that these sites are not only highly toxic, but they are also unregulated *right now* which means that without any people around to observe them it is unquestionable that these sites would have been causing havoc on the environment through which Rick leads his group but there is no mention of having to avoid a toxic coal industry hazard.

There are also pipelines strung out across the entire southeast, though, to be fair, the southeast has far fewer pipelines than other areas of the country, the Atlanta region is crisscrossed by refined oil and natural gas pipelines which, eventually, would breakdown and cause widespread destruction along the routes and in the areas where the survivors are located.[29] The group also lived within range of, or traveled incredibly close to, 12 nuclear facilities all of which would be, at least by years end, potentially producing quite a bit of radioactive activity. Which, even if the group managed to avoid the worst possible fallout scenarios evidence of these facilities breaking down would be visible to them as they traversed the area between Atlanta and Washington, D.C.[30]

There are at least thirty hazardous chemical plants in the greater Atlanta area not to mention the hundreds of such plants that punctuate the landscape along the routes Rick and his companions travel, yet, there is no indication that these plants and facilities are doing any damage to the group or the environment either directly or indirectly.[31] Many of these sites could leech hazardous chemicals into the environment or combust and, according to the EPA, which currently monitors 12,000 such facilities nationwide, 9,000 of those facilities pose a serious hazardous risk to their communities, listing them as catastrophic hazards or worst-case scenario facilities—included among those 9,000 are all 30 in the Atlanta area to say nothing of the hundreds more that line the escape route taken by the survivors to get from Atlanta to Washington, D.C.[32]

I Can See You Make a Habit of Missing the Point

Washington, D.C., would be far worse off than Atlanta as a place to take up refuge. The population of the Washington, D.C., metropolitan area, which includes the area of Alexandria where, presumably, the survivor safe haven is located that Aaron invites Rick's group to join, is 6.1 million. The amount of resources in this area is going to be vastly superior to the Atlanta area, but so will the zombie threat, by roughly a million more. Of course, it seems equally likely that there will be more survivors in the area as well, and that turns out to be the case by a substantial margin of difference. But far from the zombie threat being the most insidious danger in the Washington, D.C., area, the aforementioned pipelines, chemical and nuclear facilities, as well as coal industry threats are also greater in and around D.C. Comparing these two metropolitan locations, Atlanta and Washington, D.C., as the two long-term refuges inhabited by Rick and his companions shows that neither place is likely to be able to maintain even a minimal standard of living.

Per the Energy Justice Network, which works to map all the hazardous plants and facilities in the United States, there are 85 hazardous locations within 50 miles of Atlanta's city center. These sites are capable of destroying the environment, though to extremes of varying degrees, it stands to reason that individually these locations are dangerous enough, but when accounting for the simultaneous removal of the people that could identify, prevent, contain, or clean up such occurrences, the synergistic effect of these sites together would effectively neutralize the ability of anything, not just survivors, to live in these areas.[33] Compare the 85 such sites in a 50-mile radius of Atlanta—the known area Rick and companions, not to mention the other survivors that inhabit this area—to the 142 sites that exist within a similar radius of the center of the District of Columbia. Most of these sites are collected in and around D.C. and Baltimore and, between these two urban centers, it is entirely reasonable to expect that the Potomac River watershed and the Chesapeake Bay would be so polluted that the attempts to rebuild society here would not even be an option.[34] Of course, if this is not enough to make the D.C. area unlivable, consider that upstream, 150 miles away, sits Philadelphia. Within 100 miles of Philadelphia's city center there are an additional 749 such hazardous sites. If only one quarter of those sites exist in a southern quadrant that would have a synergistic effect on the Washington, D.C., area, then it is possible to add another 187 sites to the problems compounding in the area where the survivors take shelter in Alexandria.

Washington, however, looks like Atlanta when Rick first arrived there, only in the case of Washington, it actually looks better in some respects, and this is more than a year after the apocalypse started. Alexandria's residents, like Rick's group, are still managing to keep several vehicles running—and in the case of Rick's group they have managed to keep their two and half ton truck on the road all the way to the gates of Alexandria. That the resources, know-how, and usability of these vehicles is still factoring in to the story clearly trades on the idea that there is an abundance of resources that are not going bad, getting contaminated, and are easily locatable, none of which really makes sense almost two years into the apocalypse—especially for a group of wayward vagabonds like Rick and his companions moving through territory that is completely foreign to them.

Alexandria is set in stark contrast to the apocalyptic world right outside the gates of the community. About halfway through chapter 12, when the survivors with Rick make it to Alexandria, Kirkman offers a wide lens view of the safe haven, and there is a clear demarcation between the apocalypse, represented by broken buildings, garbage and debris, and a general sense of filthiness, and Alexandria which is whole, healthy, clean, and a functional community seemingly immune from the nightmarish conditions that exist on the outside.[35] In fact, Aaron tells Rick that within the walls of Alexandria, "for the most part, [they] are able to return to the life [they] remember."[36] Life in Alexandria centers on the social milieu, shoring up defenses against the zombie threat, and expanding the community even though there is plenty of empty space within the walls. Most of the action after the group arrives in Alexandria takes place in the artificial interior of the walled in spaces and when the group sends people out for supplies they never really encounter anything worse than the now run-of-the-mill minor dilapidation of the cityscape populated by zombies.

Soon, as is to be expected in an area as overrun with zombies as Alexandria, the zombies breach the walls and chaos ensues resulting in the deaths of several community members. The big reveal of the Alexandria sequence in the narrative is when, after the citizens have defeated the zombie threat and the breach has been contained, Rick, standing amidst the carnage, covered in gore, unleashes a long soliloquy about the revelation the fight has brought to him. Rick tells the group that he now understands that the zombies are "a manageable threat" and that he believes together they "can survive anything" and in doing so they "can rebuild the walls ... make [the] community better than it ever was" because he can "see [their] shortcomings ... and how to eliminate them" so that, for the first time in a long time, "the road ahead ... seems long and bright."[37] The ultimate message Rick delivers is that finally, after all their struggles and loss, Rick suddenly finds himself overcome with something that had long ago been lost—hope.[38]

After this revelation and renewed commitment to ushering civilization back into existence the story shifts its focus away from the zombie threat—because now it is manageable—and places society at the center. Not just the survival of a small group, but the development and extension of society across the land—Rick, and his companions, will form the nucleus of the rebirth of society less than two years after the apocalypse. The message Rick carries forward in the story is not just making Alexandria greater than it was, but the full-on re-establishment of civilization, as he confidently explains to the leadership of Alexandria after they have completed the cleanup of the zombie breach.[39] Completely ignoring that, even if they have, up to this point, managed to avoid the worst that could have happened, the worst that could happen is still in the cards. Soon, supplies start running short and, for the first time in more than two years, the survivors are starting to actually worry about finding enough canned goods to sustain them as well as other resources that they have been, apparently, finding and using with little effort.

But, as luck would have it, shortly after this realization, Jesus shows up with good news. Like Aaron before him, he brings good tidings about other communities, an offer of friendship, and, most importantly, an offer of trade. Interrogating Jesus leads Rick to the conclusion that they must throw their lot in with Jesus and his group because they cannot continue on like they have been. While discussing the possibilities that come with such an offer, Rick leads his most trusted lieutenants to a hill on the outskirts of a nearby residential area. He explains to them the importance of joining forces with Jesus's community, he talks at length about the world being safer and better if they work with other groups of survivors, and, ultimately, he leads them to the top of the hill to overlook Washington, D.C., and the whole view is serene. The sun rises in the background, illuminating the capital, and revealing it to be largely intact, primarily devoid of human life and activity, but otherwise in fine working order considering the length of time that has passed and the numerous events that have taken place, and Rick says, while staring out hopefully from his vantage point, "I think we've lost sight of what's out there on the other side [of our walls].... A larger world ... it's all around us, waiting on us to become a part of it again. We just have to be brave enough to accept it."[40]

At the close of the second compendium readers are introduced to the Hilltop community. This is a sprawling complex, hemmed in by incredibly high walls, housing almost two hundred people, with industries, food production, medical capabilities that exceed anything Rick's group has encountered outside Woodbury, and it is continuing to grow. Later in the story, one of the residents of the Hilltop will tell Maggie that sometimes she does not think her son "realizes the world has ended and [they] need to conserve [resources]."[41] The Hilltop represents the return to civilization Rick has been prophesying throughout most of the second compendium. At the end of the second compendium Rick beseeches his companions to embrace the possibilities that the Hilltop represents because, if they do, then they "can finally stop surviving and start living."[42] This call for a return to normalcy is the thread that binds the majority of the second compendium together and is, at this point, the force driving the narrative forward.

The Right Choice Is the One That Keeps Us Alive

Though the group has been dealing with increasing amounts of social destabilization throughout the story so far it has only shown up in small highly charged episodic moments. Ignoring the zombie threat, Rick's group has dealt with a serial killer (Thomas), a sociopath (the Governor), a psychopathic child (Ben), two suicide attempts (Carol, successful, and Maggie, unsuccessful), marauders, bandits, and cannibalistic threats from numerous external sources, and plenty of internal strife both before and after joining the Alexandria community. In addition to the focus on acquiring the resources necessary for survival, the social destabilization detracts from the reality of a post-apocalyptic world. The characters are often seen lamenting the loss of societal luxuries, but their real concern is how quickly they can return things to normal.

Every time *The Walking Dead* characters discuss the future as a return to normalcy what they mean—implicitly or explicitly—is a return to *social* norms, law and order, comfort, etc. The way things were in these discussions assumes a static world, an environment which, if it is out of hand, is easily re-conquerable, but which never seems out of hand, threatening, or beyond the ability of the various groups to regain control of it along the way to returning to normal. Such an assumption is almost forgivable in the early telling of the apocalyptic story, but, by the time the group stands awestruck in the center of the Hilltop community, they have crossed and crisscrossed more than 1,500 miles of the southeast United States, surviving for more than two years, and there is something amiss in the hopefulness of Rick Grimes and the seemingly easy path before them to returning to civilization.

The third compendium brings with it the hallmark of any civilized society—full scale war. What the Alexandrians come to realize is that they have very little to trade except their ability to wage war, which they do on behalf of the Hilltop, in order to secure much needed supplies. However, waging war against Negan and the Saviors is not as quick as their conflict with the Governor. The war itself is of little interest except that it takes place across a wide-ranging area around multiple communities and not once do the myriad survivors encounter an area that is environmentally off-limits to them except for those areas that have a high density of zombies. What is particularly different about this conflict is that the two primary groups engaged in the conflict are led by men that genuinely believe they are going to bring civilization back to the world. Negan tells his group, upon returning to his compound to discover there has been some disloyalty among the ranks in his absence, that the rules cannot be ignored because the rules are what keep them alive. The rules, according to Negan, "are what make everything work" and by following the rules the Saviors will "bring civilization back to the world."[43] The whole point of the conflict that plays out between Negan and Rick is a war for who's vision will determine the path back to civilization.

This is made clearer when The Kingdom is introduced, and Rick forges an alliance between the three groups to take on the Saviors. After the first major engagement, where the tri-communal alliance is exposed to Negan and his troops, Jesus tells Rick that Ezekiel means well and that he believes in what Rick is trying to do by bringing Negan down, because of all the men leading communities in the all-out war taking place, Jesus tells Rick he is "a leader [they] can follow."⁴⁴ The war exposes nothing new or different about the world except that at the height of the war for who will control the definition of civilization, it appears that the decay and collapse seems to have slowed down considerably or stopped altogether. The world is shown to have reached rock bottom and all that is left to do is determine who is going to manage the recovery. From the point The Kingdom is introduced all the characters throughout the graphic novel are genuinely committed to the idea that recovering from the apocalypse that has befallen them is merely a matter of organizing the survivors into productive groups and repopulating the planet. After Negan is defeated Rick tells him that they are now going to rebuild and "undo all the damage [he] did," and that far from executing Negan for his actions Rick is going to make him watch from a jail cell so he can see "how wrong [he was] ... how much [he was] holding [all the communities] back."⁴⁵

From the end of the war with Negan the story jumps forward *several years*. The first glimpse Kirkman provides of the communities they are bustling not only with the normal activities of growing and blossoming communities, but the groups have reached a sufficiently stable level of civilization that they feel as though they can build and put on a county fair. At this point the story arc has completely abandoned any notion that the environment was destroyed or substantially undermined by a massive human die-off, there is no longer any potential for the environment to factor into the story as an antagonist or even as an obstacle to the communities successfully returning to civilization. It may be easy to fill in the gaps on Kirkman's behalf since he has jumped so far forward and assume that the group had great difficulty in creating an orchard, or building the windmill and finding the wherewithal to have grains to make using the mill worth the cost, nor is there any reason to suspect that building and securing safe passageways between the communities was difficult beyond avoiding the places with an abundance of zombies as the joint communal effort of mapping out and clearing of new areas has become nothing more than a routine removal of zombies.

The world that is shown to the reader in the third compendium is Edenesque. Even after the group encounters the Whisperers, Carl is told by Lydia, a captive taken from the Whisperers, that the Whisperers survive by finding "berries or gardens that have grown wild, fruit and other things. [They] also hunt. There are great herds of animals [they] follow sometimes ... [and] sometimes the dead kill an animal and [they] share that."⁴⁶ In the depictions of Alexandria, the Hilltop, and the report made by Lydia to Carl as well as, later, The Kingdom and Oceania, a fishing community now linked to the others, the world is just fine—full of resources and as compliant as it has ever been, even to societies operating with the most primitive of technologies.

Herein lies the link to the hopefulness that dystopian apocalyptic narratives build in their presentation of the world. It is not enough that the survivors managed to avoid the worst pitfalls a near total pathogenic destruction of the population would likely entail, but the environment provides a cornucopia of resources and opportunities to the hardworking and intrepid survivors rewarding their ingenuity with abundance and comfort. This representation of the apocalyptic world minimizes the pointless suffering that usually accompanies large scale disasters—never mind a global disaster.

The Walking Dead frequently offers views that do not corroborate what we know the world should look like and, as the story drags on, these depictions become disorienting to a proper appreciation of just how tenuous our current situation is and how bad it could become. Consider the end of the third compendium when Rick and a few fellow companions are searching for a runaway Carl. Upon their return to the Hilltop they encounter a grisly border established for them by Alpha, the leader of the Whisperers. While it is certainly the case that the horror of a fence made from severed heads is quite the distraction, it only takes a moments reflection to realize that the search party is traveling in an area their communities have not mapped nor cleared. It is an area that is currently controlled by a group of people who, for the most part, choose to live a Spartan existence on par with animals, consciously choosing to make very few if any improvements to the land surrounding the area they inhabit because the old world is dead. And yet, the grass in this field, near a residential area, we know has received no maintenance *for years*, is not even ankle high to the people standing in it.[47]

Of course, of all the aspects of the world Kirkman's story gets wrong about the post-apocalyptic world, it is the absence of animals that is most glaring. Not just the domesticated animals the communities in the third compendium have managed to corral and maintain for the use of the survivors. But the animal populations that would be able to flourish in this world. After all, if the people in the story are inhabiting a world that is generally non-threatening and full of abundance, why should the animal populations fair worse? If something as deadly as the zombie virus envisioned by Kirkman struck human civilization, even if it did not create zombies, it would leave the world blanketed by dead bodies.

With so much easy food lying around large predator species and scavengers would see population booms very quickly—especially with so few humans left to hunt them. Driven by the necessity to hunt most of the survivors would lack the skills to take large game animals, especially once ammunition became scarce, and as a commodity, in a world where the rule of survival is "fight the dead, fear the living," the use of ammunition to hunt would be wasteful as long as successful scavenging was possible. Taking for granted, then, that the animals capable of stalking and killing humans would be on the prowl it is strange that no one has come face to face with a cougar or a bear or wolves despite near continual scavenging and movement through areas that would likely be attracting these very animals to say nothing of already being home to these animals.

Again, this is assuming no one threw open the gates to the zoos as people were dying in droves. In *The Walking Dead*, Ezekiel reveals to Michonne that he was a zookeeper before the apocalypse started. He claims that when he made his way to the zoo most of the animals left "were trapped, starving, and all alone..." but he does not reveal if he freed those animals or just saved the tiger he keeps as a pet.[48] It seems fair to presume that if he was concerned at all for those animals he would have let them out of their cages. It is also difficult to determine if being one of the last animals left indicates that the animals had all died, or if the majority of them had been freed already. What can be known is that it is entirely possible that animal loving people all over the country may have set those trapped animals free and those animals would have seamlessly integrated into the chaos of the collapsing world. There are 350 or more zoos in the United States among the nearly 2,500 animal exhibitors licensed by the Department of Agriculture, housing 750,000 animals across some 6,200 different species. Obviously, some of these animals are non-native and their ability to live and thrive in a foreign environment would raise significant challenges to their success. Still, if even half of those animals were released there is a good chance that the fauna maneuvering through the world would take on a different dimension—one that would, at some point, be forced into interaction with people traipsing about the countryside.[49]

It is exactly this kind of fanciful presentation of the apocalyptic world that creates genuine misconceptions about post-apocalyptic life. To be a survivor in the aftermath of Kirkman's zombie apocalypse is to be an inheritor of the earth but not a planet that is covered by a rapidly imploding environmental situation. It is to be given the great opportunity to start from square one and rebuild society correctly, that is, as a utopian construct—which is exactly what Rick Grimes intends to do, it is the very basis of his story arc. This very notion, rather than prompting us to seriously consider the state of the world and what could happen to it—and to us—instead trades on the fetishization of victimhood "to which the only indisputably ethical gesture [is consenting] to one's own exploitation in the name of a humanitarian ideal."[50]

This very idea is put into words by Rick in a conversation he has with Andrea about Negan toward the end of compendium three. When pressed why they do not just kill Negan, Rick asks Andrea if she really gets what is going on, what is at stake, he says, if I'm going to lead these people they need to respect me. They need to look up to me. They need to see me as more capable, not better, but more capable than they are. To a certain extent, I'm what holds this place together. Killing Negan is the expected thing, killing Negan is what everyone wants, I'm the one who doesn't kill. I'm the one who says there is a better way and that makes me a leader. I'm doing the right thing instead of the easy thing. But more than that I'm showing them that we're better than our emotions, we're more than our rage and fury, our anger and hatred. We're civilized people ... if we ever lose that ... that's when [everything] starts to fall apart.[51]

When this idea—I'm doing it for humanity—is coupled with our socio-ontological commitment to the 'frontier,' and no longer having a last frontier for people to cross, the post-apocalyptic landscape, far from being deadly and foreboding, becomes desirable, thus, fueling the skeptics beliefs that things will not be too bad.[52] Not only will things not get too bad, but even when they are at their worst, we will not only survive, but rebuild civilization better, and Rick is no longer a fictional folk hero he is the standard for hoping that the apocalypse happens. Rick Grimes provides a lens to witness the present status quo devolving into apocalyptic chaos and to watch it become desirable as a way to remove the socio-cultural obstacles to ushering in a better future.

Don't Let This World Spoil You

There is a certain moral slippage that inevitably accompanies life in a post-apocalyptic world. It may be why the post-apocalyptic is so unsettling, made evident in the early struggles of the survivors to define what is right and wrong and socially permissible, but it is definitely a driving force behind the hopefulness of the story arcs that develop out of such a narrative structure. The return to "civilization" means a return to moral imposition. Not just the return to moral behavior, but the sense of fostering a social system that can impose rules for moral behavior as well as a means for delivering punishments when the rules are broken. The return to civilization promised by the central leaders is reflective of the notion that morality and virtue can be codified and enforced which, more than anything else, indicates a desire to return to what was rather than *actually* ushering in a better, improved society. What marks the dysto-apocalyptic world out as particularly special with respect to hope for a better future is just that—the recognition that we could, in a post-apocalyptic world, not just return to civilization, but that we could improve it, we could make it *better*. The failing of Rick's vision for the future is that it fails to encompass the possibility of a better future opting instead for an increasingly close approximation of what the world was like pre-apocalypse.

The questions raised in the context of *The Walking Dead* about societal norms, social destabilization, political structures, morality, and many other human-centric concerns are perplexing and rife with opportunities for philosophical discussion. However, they are raised in a context that foregrounds those concerns as if the world just stopped working and it ought to be clear—especially in the case of *The Walking Dead*—that such a premise is as far from reality as one can get. Nevertheless, following Peter Sloterdijk, the zombie is the catastrophic ideal for discussing the breakdown of society precisely because zombies possess a face but lack human-ness. Their ruin represents the ruin of the social order where the social is the location of properly human personhood.[53]

An obvious challenge to this criticism is that Kirkman's story is just that, a story, and as such it is exempted from a need to present the world as it would *actually be*, after all, there could not, presumably, be zombies either. Well, we think not, but one mad scientist, some genetic manipulation of a virus, a little luck, and voilà, maybe there are zombies; but, regardless of whether or not zombies would be joining us in the post-apocalyptic world, the world would fall into disarray in predictable ways if a biological pandemic caused massive, rapid population decreases worldwide. The way the world falls apart, and how survivors, like Rick Grimes, choose to live in the aftermath, is precisely why we ought to expect the world to be accurately represented—the world is the context for the all the decisions one would have to make about societal structure, social norms, and morality not to mention whether or not, and if so, for how long, survivors would be able to live at all, never mind the level of comfort that Kirkman seems to afford his characters.

The federal government has long been in the business of preparing for large scale disaster scenarios going so far as tongue-in-cheek preparations for a zombie apocalypse because, as the saying goes, if you are prepared for the zombie apocalypse, then you are prepared for anything. But this is clearly wrong if the zombie apocalypse looks anything at all like the one Kirkman has envisioned. The operating premise behind this claim seems to be that the zombie apocalypse will happen with the same plodding rapidity of global climate shift, or that it will be localized like a nuclear event. These assumptions are wrong because they too narrowly conceive of the aftereffects of the zombie apocalypse as being precisely what Kirkman has made them out to be. As such, it is only fair to claim that in the event of a zombie apocalypse—or any near total mass die off of the human population—things would not be manageable, or livable, or easily recoverable. This leads us to the obvious conclusion that, perhaps, Kirkman's zombie narrative cannot be the gold standard for pathogenic apocalyptic storytelling. So, let us turn to a story that focuses on a more reasonable pathogen caused massive die-off, but does so without going for near total mortality.

4

Pathogenic Shaped Futures, Part II

Reduction and *Y: The Last Man*

I'm Not Afraid of the World

Y: The Last Man[1] deals with a storyline somewhat similar to Kirkman's zombie tale, but, rather than have their virus kill off a near total percentage of the population, Brian K. Vaughn and Pia Guerra instead chose to kill off the entire male population of every species on the entire planet, save two individuals, Yorick Brown and his pet capuchin monkey, Ampersand. The plague in *The Last Man* differs from the plague in *The Walking Dead* in two critical ways: one, the dead stay dead in *The Last Man* much like they would in real plague conditions and, two, everyone that dies as a result of the virus in *The Last Man* dies simultaneously and instantaneously whereas Kirkman's virus spreads quickly from a patient zero to everyone else. How an apocalyptic pathogen operates is of singular importance to understanding how life in the aftermath of such mass death would be able to continue.

More than just reducing the global population by half, however, the story of the last man is global in scope, with the story kicking off in Israel, the Australian Outback, Washington, D.C., and New York City, simultaneously creating the character links that will bind the story together from the start. Of course, the story itself follows the main character, Yorick, as he travels the globe in the aftermath of the "gendercide" and, subsequently, opens most of the globe to an apocalyptic examination that spans both time and geography. Also, though it seems somewhat more mundane than other points of departure from *The Walking Dead*, Vaughn and Guerra render their story in full color which makes it more difficult for their narrative to bury the telltale signs of the apocalypse in the grayscale of our imaginations. Being able to see quite clearly that the panoramic vistas portrayed in *The Last Man* bear very few, and quite often none, of the scars of an apocalyptic event gives

Vaughn's story less of a margin for error in the representational accuracy of post-apocalyptic life.

One final, though important, difference between the two stories is the role science plays in developing the plague narrative. In Kirkman's zombie apocalypse a scientific explanation for the viral destruction of human life is limited to the carefully crafted lie Eugene uses as a façade to garner protection from Abraham and respect amongst the survivors. Once Eugene is exposed as a fraud there is no more mention of the cause of the zombie outbreak that allude to causation or a deeper understanding of the virus. The survivor's in *The Walking Dead* appear hopeful of a rescue, most vehemently in the conversations that Rick and Shane have about moving the first camp. Some of them, especially Hershel, are hopeful a cure will be found though there is no inkling of how that might come about, or even if it is possible given how quickly and indiscriminately the virus felled most of humankind. The disease does spark fear amongst the survivors, and not just because they do not want to become zombies themselves, but, for instance, during Lori's pregnancy the survivors are fearful of what might happen should she miscarry the child—the obvious and most odious assumption being that the half-formed fetus would eat its way out of her womb.[2]

Because they do not understand the plague none of the characters is positioned to shed light on the causes, possible cures, or any deeper understanding of the virus beyond what they can glean from common sense experience. And, since the majority of the characters are not doctors or scientists, what they are capable of gleaning is largely unhelpful for overcoming the virus though it does help them manage life in the post-apocalyptic world—the few doctors that do exist in the narrative are killed off rather quickly and none of them engages in any scientific study of the virus. The one lone exception to this is Eugene who pauses to consider a zombie specimen that cannot muster the strength to attack the survivors while they are gathering supplies. So, the height of scientific engagement with the virus in *The Walking Dead* amounts to Rick deducing that every living person is already infected, the survivors accurate determination that the bite of a zombie is what kills you no matter how much poisoned gore you get on you while slaying the zombie hordes, and, through trial and error, they are able to figure out that, if they work quickly enough, it is possible to save a life post-bite if the injured area can be amputated.

In *The Last Man*, however, one of the central characters, Dr. Alison Mann, is a scientist, and a bio-geneticist—working on cloning no less—which gives both the other characters and readers a portal to access a deeper understanding of what is going on and how to overcome the virus. Of course, it is the character, capabilities, and actions of Dr. Mann that makes *The Last Man* a hopeful dystopia. The narrative is developed around Dr. Mann's abilities to determine the cause of the gendercide, develop a cure for it to protect future males, and, to ensure that there might be future males, successfully clone Yorick.

The centrality of science to the story seems to be a necessity given that half the world's population is still alive, and many of those people would be scientists frantically working to avoid the extinction of life; but also, the scale of the viral destruction, the sheer thoroughness of the elimination of males, makes it impossible that there should be any solution *other than* figuring out how to clone Yorick Brown, thereby preserving the immunity to the virus genetically. Yorick, though the main character, is actually present to be the petri dish for Dr. Mann's experimentation to solve the riddle of the gendercide.

Another in a Long Line of Emotionally Crippling Misadventures

The Last Man is, however, an apocalyptic tale and the total, simultaneous, and wholly unexpected destruction of the male gender means that the world post-gendercide would be utter chaos. Despite the survival of a large number of people society almost immediately fractures between those attempting to maintain order and rebuild, those attempting to build something new, and the unencumbered now searching for purpose and direction in a world untethered from the demands of globalizing capitalist industrial patriarchy. While certain functions of society are underway to normalize the situation—cleaning up the billions of dead bodies lying about—many other social functions have been abandoned or are nearly impossible to operate efficiently due to insufficient knowledge, low prioritization, or lack of a labor force. Before unpacking these various societal obstacles, however, it will be helpful to put the scope and scale of the gendercide into perspective.

The Last Man represents a vision of the world that is "unmanned" and rather than verify the destruction as presented by Vaughn and Guerra, I will rely on their statistics because it is their statistics that provide the context for the world that Yorick inhabits. Obviously, there are certain aspects of their world that will require interrogation, or extended analysis beyond what is provided, but it seems salient to allow their representation of the destruction to take center stage. As such, the beginning of the graphic novel provides a fairly horrifying visual experience of the apocalypse. "Something's wrong!" So begins the story of the last man on earth as a woman covered in blood seeks help for her dying sons. In the frames that follow a ticking clock sequence of events is shown leading up to the moment of gendercide.

Yorick, in Brooklyn, is on the phone with his girlfriend, who is studying abroad in the Australian Outback. Yorick's mother, a congresswoman, is having a heated conversation with a senator about legislation and abortion. Alter, a commander in the Israeli Defense Force, is engaging with Palestinians while talking to a news crew in the West Bank. Agent 355, a United States super-secret service agent, is attempting to recover a sacred and culturally sensitive artifact in Jordan. Dr. Alison Mann is in the throes of a premature and unnaturally painful birth in Boston as Yorick's sister, Hero, is having an afternoon on-the-job tryst across town. Each frame is preceded by an announcement of dwindling minutes leading up to the unmanning of the world. At that critical moment a series of visuals from across the world are provided so that not only do we see Congresswoman Brown's assistant die in her office, but the all-male news crew in the West Bank collapses, Agent 355's pilot dies mid-flight, and Hero's boyfriend hemorrhages in her arms.[3]

At the same time, however, we are given a glimpse of the body strewn floor of the Tokyo Stock Exchange, a blood covered priest in the arms of a nun at the Vatican, a small girl holding her dead puppy yelling for her father (who we know must also be dead), two prostitutes standing over a bloody John in the streets of Amsterdam.[4] A referee dead at midfield of a women's soccer match in Sao Paulo, a lone panic-stricken woman surrounded by her dead colleagues at the Johnson Space Center in Texas, a screaming alarm at a nuclear facility in Leningrad warning the unresponsive dead of a pending emergency, and, finally, a person-less frame containing a train wreck and a dead giraffe in Mombasa, Kenya.[5] The carnage is totalizing in such a way that it is clear the world is now in a state of chaos and the ensuing destruction, caused by the immediate cessation of every activity men were performing, without warning, is crystallized by a New York police officer just before she commits suicide: "All of the men are dead."[6]

These depictions of the bloody, messy end of the males worldwide is punctuated by a series of smaller frames showcasing the panic, fear, helplessness, and confusion of the surviving women that comprise the primary cast of characters in the story. It is at this point in the story, once it has been established that all the males globally are dead, simultaneously casting the world into disarray, that Vaughn and Guerra tip their hand and clarify just how damning the gendercide apocalypse would be to a world populated by women. Welcome to the unmanned world, Vaughn and Guerra announce, before explaining at length that

> a plague of unknown origin destroyed every last sperm, fetus, and fully developed mammal with a Y chromosome—with the apparent exception of one young man and his pet, a male capuchin monkey ... this "gendercide" instantaneously exterminated 48% of the global population, or approximately 2.9 billion men. 495 of the Fortune 500 CEO's are now dead, as are 99% of the world's landowners.... In the United States alone, more than 95% of all commercial pilots, truck drivers, and ships captains died ... as did 92% of all violent felons. Internationally, 99% of all mechanics, electricians,

and construction workers are now deceased ... though 51% of the planets agricultural labor force is still alive ... 14 nations, including Spain and Germany, have women soldiers who have served in ground combat units. *None* of the United States' nearly 200,000 female troops have ever participated in ground combat. Australia, Norway, and Sweden are the only countries that have women serving on board submarines.... In Israel, all women between the ages of 18 and 26 have performed compulsory military service in the Israeli Defense Force for at least one year and nine months. Before the Plague, at least three Palestinian suicide bombers had been women.... Worldwide, 85% of all government representatives are now dead ... as are 100% of Catholic priests, Muslim imams, and Orthodox Jewish rabbis."[7]

Clearly the number of dead represented here is meant to minimize the collateral deaths that would have occurred and ended the lives of countless women who were at the mercy of men at work. For instance, we know that Agent 355 is in a plane piloted by a man who dies at the controls. She is fortunate to have been able to take over and save herself, even if just barely, but how many planes were airborne that had no one in a position to avert a disaster? According to the statistics used by Vaughn, 95 percent of the world's pilots, freighters, and maritime captains perish. To be able to determine a rough estimate of the amount of damage such a loss would create it will be instructive to consider them as separate entities.

This Is Officially the Weirdest Nightmare I've Ever Had

At any given moment, there are an estimated 5,000 planes flying in U.S. airspace.[8] Not all of them are commercial, a specification that Vaughn provides, but if you adjust for non-commercial flights, military flights, and personal use planes, it seems reasonable to push that number upwards by at least half. So, given 7,500 planes in the air, and considering that 95 percent of the pilots are men, a fact confirmed by the Air Line Pilots Association, means that only 2,650 members of the ALPA are women. The numbers are worse internationally, as only 4,000 women are licensed to fly of more than 130,000 pilots according to the International Society of Women Airline Pilots.[9] Generously assuming all 2,650 female pilots are at the helm of airborne planes at the time of the gendercide—and that none of them are flying together—there would be 4,850 planes falling out of the sky. Estimating an average of 100 passengers per plane, a number far below the capacity of a Boeing 747, but substantially larger than the 3-passenger capacity of a Cessna 172, there would be roughly half a million people on the planes falling from the sky.

Of course, many of those passengers would already be dead, but the women on those planes would likely perish in the carnage of the crashes. If the ratio of men to women were to mirror the ratio of men to women in the global population, then approximately half of the passengers in the air would be women, or roughly 243 thousand people—just in the skies above the United States. Internationally, allowing for twice as many flights globally as there are in United States, and far fewer women pilots, the death toll would be even higher. Ten thousand flights, again allowing for each of the 3,000 female pilots to be airborne with a male counterpart, there would still be 7,000 planes dropping from the clouds worldwide. Holding the 100 passenger estimate steady, 700,000 passengers would be plummeting to their deaths—or, the females would be, and that would be 350,000 or so people. Totaling approximately 600,000 women that would perish in the immediate aftermath of the apocalyptic gendercide.

That is an additional half million dead people, which does not seem like a lot when 3.5 billion women make up half the global population, but those losses would include a lot of skilled labor that would be essential to the survival of the human race and civil society. Those that managed to survive the plane crashes would most likely succumb to wounds, exposure, or shock as there would likely be no one to respond to the wrecks. It is the wrecks that are a bigger concern, really, than the number of women that would die in the wreckage. Each plane, the majority of them flying in and around urban areas, would be falling into areas that would simultaneously be falling victim to panic not to mention rural areas with too few people to organize a proper response. Consider just one major population center: Chicago, the third largest metropolitan area in the United States, home to the third largest and fourth busiest airport in the country.

There are eight airports in a fifty-mile radius of Chicago, one of which is the international hub O'Hare. Chicagoland has a metropolitan population of 9.4 million people—half of whom would be dead before the first plane hit the ground. O'hare International Airport handles approximately 2,350 flights a day which includes carrier, cargo, commuter, and other forms of aviation.[10] This, of course, does not bear a realistic resemblance to the number of planes that would be in the air above and around Chicago because there are seven other airports. But, if one-third of the flights handled by O'Hare were in the air at the time of the gendercide there would be 780 flights within the metro area *just coming to or leaving from O'Hare*. If the other seven, smaller airports are responsible for a quarter of the flights that O'Hare handles, then the number of planes falling to the earth would number roughly 2,150. Two thousand one hundred and fifty planes crashing into an urban area with 4.7 million people still alive. The destruction of Chicago would be thorough, but the real question is how many of those women on the ground would be maimed or killed by the planes crashing into the city and surrounding areas?

Bear in mind that this is just considering Chicago. Every other major metropolitan area would be similarly pummeled by aircraft hurtling back to earth and lots of urban areas with smaller airports, or residential areas unfortunately located directly under flight paths, would find themselves bombarded in the minutes after the gendercide. Each plane would have the potential to start fires, collide with hospitals, bridges, or other important infrastructure, the death toll would be considerably higher than just the men and the aftermath would be a lot less clean than it is presented by Vaughn throughout the story. However, Vaughn not only misses an opportunity to reflect the horrific reality of mass airplane crashes the three scenes that depict plane crashes bear discussion to demonstrate just how off the mark Vaughn and Guerra are in this regard.

At one point in the story, as Yorick is aboard a train leaving Boston, the train passes a plane crash site. The plane, a single propeller Cessna-style civil utility plane, looks like it only had a bumpy landing, but it is easy to get the impression that everyone walked away from the crash.[11] In fact, the crash dots the landscape like an apocalyptic afterthought rather than a product of a massive depopulation event. The airplane issue is treated substantially only twice more during the story. Both instances are flashbacks, however, with one detailing the chaos of the gendercide in-flight and, the second, a more substantial depiction of a plane wreck that Yorick and his companions stumble upon during their travels.

In the former instance Yorick is swapping personal stories with Beth, a woman living in an abandoned church in California, and she explains the scar on her face is from surviving a plane crash. She tells Yorick she was working as a flight attendant on a plane traveling from Boston's Logan International to Los Angeles's LAX and in her recounting of the experience she conveys the panic and fear of finding herself the only person in a position to attempt to land the plane or at least divert a major disaster. Unable to land the plane, first with assistance from air control, and second, when she realizes the pilot activated a flight assistance program, she says the plane was at "10,000 and dropping hard," which is unsurprising.[12] Moments later the plane crashes and Vaughn and Guerra give a whole page to the impact scene. An open field in the middle of nowhere and, though the plane buckles from the impact and there is a minor explosion as a wing breaks away, we already know that there were survivors.[13]

The latter instance is far less dramatic, as Yorick and company come across long abandoned wreckage also located in a field. Knowing how interconnected and nuanced the storylines are the likelihood that this is Beth's plane is heightened by the fact that the location of the plane is not revealed and the wreckage is in a field with a broken wing and similarly damaged fuselage. In all fairness, the lone dead body shown in detail does not appear to be desiccated or picked over by animals, so it is also likely that the wreckage is discovered earlier in their trip, somewhere on the Missouri to Kansas leg, perhaps even earlier. However, there is nothing to suggest that the plane burned nor is there any evidence to suggest that some of the dismantling of the plane was not done *ex post facto* by survivors scavenging for useful items or materials. More importantly, the field is not damaged, there is no furrow cut in the dirt, no impact crater around the wreckage, the field is green grass and blue skies, and the plane could just as easily be a thousand-year-old archeology find as an artifact of a global apocalyptic event.[14]

Good Thing Our Shit Always Goes According to Plan

Falling planes would only be the most terrifying feature in the minutes after the gendercide. In the United States, there are 253 million cars "on the road" but obviously not at the same time. Certainly, the number of cars on the roads at any time depends on the day of the week, time of day, time of year, and other factors to be sure, but if one-eighth of those cars were on the road at the time of the gendercide that would mean a fair estimate of 31.7 million cars would be traveling the roads, byways, highways, and interstates across the country. If half of those cars were being driven by men, then 15.8 million cars would simultaneously wreck. This number does not include Vaughn's "95% of truck drivers" by which he means tractor trailers.

There are an additional 3.5 million registered professional truck drivers in the United States and 3.325 million of them are men. If only half of them were driving at the time of the gendercide, in addition to the 15.8 million passenger vehicles wrecking, there would be 1.7 million semi-trucks spinning out of control. Not only would this amount of wreckage choke off every major roadway, but the other vehicles on the road being driven by women would be colliding with many of the dead men's cars. Again, the death toll would be phenomenal. Not only deaths from impact, but the inability of any life-saving services to reach the victims that did not die immediately means that most of the automobile accidents, like the scattered survivors of the plane crashes, would die a slower and more painful, but equally certain death.

The death toll on the roads does not account for the further damage that would be created by so many simultaneous accidents. Fire hydrants knocked over, buildings impacted, traffic jams, pedestrians hurt, gasoline and other fluids that would be leaking out and igniting, and, where the trucks are concerned, chemical spills, hazardous waste spills, combustible liquids, not to mention the hundreds of thousands of gallons of diesel fuel, all of which would make the immediate aftermath of the apocalyptic event more dangerous and deadly than just a bunch of men dying. In addition to tractor trailers there would be thousands of buses and construction equipment that would also contribute to the increasing death toll and physical destruction in and around cities and along the roadways.

Finally, Vaughn includes ships captains in his 95 percent remark. If planes are the equivalent of apocalyptic hand grenades, then the ships at sea would be the ticking time bombs. There are currently more than 51,000 ships in the global merchant fleet. This number does not account for military vessels, privately owned vessels, or commercial vessels employed in local industry such as fishing vessels working as part of a local industry rather than an industrial agent.[15] If there were only half that number on the water when the gendercide occurred, there would be 26,000 ships on the water. Using Vaughn's 95 percent benchmark, that means roughly 24,000 vessels would be adrift with no one on board to safely captain the ship. Some of these vessels would drift for, potentially, years before colliding with something, and some would certainly sink, but the real question is what to do with the ships transporting oil, liquid gas products, food, or other cargoes that would spell doom in one way or another for the survivors of the gendercide.

If 100 of the ships at sea at the time of the apocalypse were oil tankers, then it is safe to postulate, using the average capacity of an oil tanker, judiciously approximated at 750,000 barrels, that there would be 75 million barrels of oil at risk in an instant. To clarify that number, there would be 3.15 billion gallons of oil drifting, unmanned, waiting to be spilt at some point in the aftermath of the apocalypse. This is just oil, not other hazardous wastes, to say nothing of the damage these ships would do if they made landfall in seaside towns and cities. Nor does this account for the passenger cruise ships at sea which would likely become floating tombs or, at best, floundering vessels that would spill their human cargo into the oceans in a frightening version of everyone for themselves as people scrambled for safety if they were lucky enough to be within swimming distance of land.

Just extrapolating these three industries with immensely high rates of men that would be killed off it is clear the actual death toll, both immediately, and for quite a long time, after the apocalyptic event, would be substantially higher than the figure provided by Vaughn identifying the total loss of males worldwide. His total does not account for the number of women that would be isolated, abandoned, or otherwise jeopardized in hospitals and nursing homes across the globe. It stands to reason that if the upper echelon of most Fortune 500 companies was eradicated those businesses would likely see production slow to a crawl or grind to a halt. There are currently 27 women at the helm of Fortune 500 companies, and though they cut across all the major sectors of business they only represent 5.4 percent of the S&P 500 list.[16] This means that the ability of corporations to respond to the apocalyptic crisis would be virtually null. Not only would these firms have lost their leadership and management elements, but a substantial portion of their work force would have been eliminated as well, taking with them years of accumulated knowledge and specialized skill sets.

Per the World Bank, the labor force of the United States was 180,800,969 in 2014.[17] However, unlike the total population of the United States, the labor force in the United States is disproportionately male as 73 percent of those employed in the United States are men, or roughly 132 million. Assuming they all survive, this leaves about 48 million female laborers that may, but, thanks largely to the asymmetrical and negative treatment of women in the work force, likely will not, possess the skills necessary to organize, implement, and manage recovery efforts—not to mention the problems facing a fragmented manufacturing, delivery, and market system that would occur as the communication infrastructure began to breakdown. By the time planes had stopped crashing, cars had stopped wrecking, and boats had stopped running aground or sinking at sea, so much collateral damage would have occurred that any large-scale efforts to stabilize society would be doomed for some time and even some local efforts would be non-starters.

It may not seem pertinent to the overall story arc of the Vaughn's tale, but when you consider the tremendous potential problems that communities would face from waste treatment facilities, backed up sewer systems, a loss of utilities, and food shortages all of which would be severely undermined by the loss of both leadership and labor—not to mention medicines, security issues, or other industrial components of First World cosmopolitanism, the apocalypse that would result from the gendercide would be several orders of magnitude worse than Vaughn and Guerra depict. For instance, Vaughn mentions military involvement as an aspect of the gendercide because most warmongering is done by men, and though warfare may drop dramatically, it would not stop and, in some ways, it may worsen, especially around suddenly precious resources like potable water. On a smaller scale, what Vaughn does not include in his statistics, is that many police forces across the country would be similarly decimated.

Using the most recent data from the Bureau of Justice Statistics and the Department of Justice, "in 2008, across 62 reporting federal law enforcement agencies there were about 90,000 sworn officers, of whom approximately 18,200 (20%) were women."[18] Of course, these officers are not spread evenly across the countries many departments as the majority of these officers are located in larger metropolitan areas and federal agencies. Even allowing that not a single officer perished in the aftermath of the apocalypse there would be far too few people to police the many instances of tribalism, looting, or violence that would surely follow on the heels of an annihilative event. Those that could would have too few resources to do it effectively meaning that there is greater likelihood that most of the world would slip rather quickly into something like the Wild West which would only add to the difficulties facing those that wanted to attempt to return society to something stable and productive. Of course, we know that at least one female officer does not survive the apocalypse in Vaughn's story, committing suicide almost immediately after realizing what had happened and we can be fairly certain that many other female officers perished in the car accidents and other collateral instances that would occur during the gendercide.

I Met This Scary Bring-Out-Your-Dead Chick Today

The world Yorick inhabits after the apocalypse should be a sprawling cesspool of destruction and waste. Given the amount of surface level damage that would exist in urban and developed areas it would not take long for drainage systems to become clogged. Clogged drainage systems would cause sewers to back up into streets and homes which, even without the mosquitos and other animals drawn to such things, would make life in these places more difficult. In addition to the wreckage from automobiles, fires both natural and unnatural in origin, conflicts, storms, and other things, the standing water and sewage would create hazards that would make scavenging difficult. Power outages would render frozen refrigerated goods inedible in days as well as potentially affecting unattended aspects of various industries. Yorick should expect to find a world covered in muck and ash piled up with debris and dead bodies. But once Vaughn provides his statistical analysis of the post-gendercide world the reality Yorick faces is quite the opposite.

Making his way from his home in New York to Washington, D.C., takes Yorick two months. A time period unaddressed by Vaughn as he goes from all hell breaking loose to Yorick walking down a sidewalk in a D.C. suburb.[19] If you were to remove Yorick from the picture it is likely that the scene would be interpreted as a quiet night in an otherwise mundane D.C. neighborhood where the neighborhood drunk has, once again, parked his car a little wonky in front of his house. It is possible that this neighborhood has managed to escape most of the devastation, or that it was inhabited by some well-organized and highly motivated women who have cleaned and protected the neighborhood—it is just as likely that one of these women is responsible for the bad parking job.

In the ensuing scenes either of these interpretations could be legitimate as the action focuses on Yorick's run-in with Waverly, a model-turned-one-person-body-cleanup crew. Waverly reveals a few key aspects of post-apocalyptic life in D.C., namely that RFK stadium has been transformed into a massive crematorium where each body turned in merits a can of food in payment. This is obviously a government run program to get the dead, decaying bodies removed from the public before they cause pandemics bringing plague upon the female survivors. This would seem especially important since dead bodies are most likely to cause pandemics of gastroenteritis if they pollute water sources as they decay.[20] However, though the focus here is entirely on dead humans, the gendercide kills the males of all species which means the clean up of just the humans would not necessarily alleviate the actual problem posed by the amount of dead bodies.

In a world where standing water, urban runoff, and waste treatment are already compromised, lingering dead bodies would likely make life everywhere more difficult. Given that Yorick is two months into his trek and Waverly has a dump truck full of bodies it is almost a certainty that the bodies that have not made it to the crematorium are well into the final stages of putrefaction. The bodies lying about at this point would be starting to split open and spill their contents. Waverly's truck is covered in a bloody sheen of gore, but the streets are clean of all debris and gore and look relatively normal, especially as Yorick makes a mad dash toward the capitol.[21]

As Yorick is making his way into D.C. to rendezvous with his mother, Agent 355 is bringing the lawful president to the White House. As she is making her way into the city the car she is using runs out of gas. She and the president, stranded on an overpass, discuss the possibilities of walking into the city. The scene is a far cry from a totalizing sense of destruction that ought to be playing out. Agent 355 laments the fact that the gendercide happened at rush hour, which means, for this analysis, there would have been a lot more vehicles on the road than the minimum assumed above—at least up and down the Eastern Seaboard. The result should be a greater degree of carnage in and around the roadway corridors. Yet, looking out over the scene that Agent 355 and the soon-to-be president are surveying all the cars are sequestered in between the guardrails. One car has hit a street light, but not knocked it over, another has hit the guardrail, but not broken through it, there are overturned cars, and one of them is leaking a strange green liquid, but no fires or evidence of fire, no collecting pools of fluids from the numerous vehicles, and in all reality the whole thing could be a run-of-the-mill 20-car fender bender pileup on a crowded rush hour interstate—a far cry from apocalyptic chaos.[22]

These two post-apocalyptic scenes frame the way readers are expected to understand the aftermath of the gendercide. That the capitol grounds are clean is to be expected if the capitol was able to escape the immediate aftermath unscathed. Which seems somewhat unlikely, given that there are two major airports servicing the D.C. metropolitan area, Dulles and Ronald Reagan, not to mention the Thurgood Marshall International Airport in Baltimore. There would have been a lot of planes and cars coming to an immediate halt in and around D.C. Even so, the White House and the capitol grounds could have escaped given the airspace restrictions in place to protect those areas. Hence, neither Yorick's encounter with Waverly nor Agent 355's arduous attempt to bring the president back to D.C. from her home in Arlington are quite in line with the reality that D.C. should be facing—even, and perhaps especially, two months after the gendercide.

This is made even clearer when considering the antagonism taking place between the remaining democrats, Yorick's mother among them, and the surviving Republicans. The latter allege a coup is underway, the former arguing that the government has largely ground to a halt—a point punctuated by the fact that it has taken two months to locate the highest-ranking female and bring her back to the White House to assume her responsibilities as president while special elections are still being planned. Amidst the political chaos playing out on Capitol Hill, Yorick leaves the White House under the protection of Agent 355 to find Dr. Alison Mann so that she can, hopefully, use him as a specimen to produce a cure for the plague. As Yorick and Agent 355 head toward Boston in search of Dr. Mann, we again see a bird's eye view of the D.C. cityscape and it continues to reflect a less-than-destructive apocalyptic event.

Fat Lot of Good Our Tits Do Us Now, Right?

At this point in the story Vaughn introduces an unexpected sense of normalcy into the narrative. In addition to the healthy, unbroken and undamaged trees lining the streets, and ignoring the fact that the streets are at least as clean as they were pre-apocalypse, the grass is still manicured looking, though maybe it is starting to grow taller than usual, the major action item here is the revelation that there are three distinct organization efforts taking place. This would seem normal if it were not for the incredibly dire situation that these surviving women should be facing. First, it has already been exposed that there are factions of women that want to maintain social standards and return society to its pre-apocalyptic functions. The women running the government want to hold elections, maybe tossing out the constitution beforehand, but largely they see the world has having been abruptly interrupted and hope to put things back to the way they were.

Secondly, there are factions of women that want to create an entirely new social order, the largest and most organized of these calls themselves Amazons. The Amazons want to eradicate the last vestiges of patriarchy by removing all traces of masculinity. The Amazons are briefly mentioned in the exchange between Yorick and Waverly, but as Yorick is leaving D.C., making a pit stop at the Washington Monument-cum-shrine to the dead, the Amazons are revealed in a panel that immediately conjures up a biker gang. The leader of the Amazons, Victoria, is adamantly opposed to the goals and ideals of the now female led government and the women who support them in their attempts to return society to anything that resembles a pre-apocalyptic societal orientation. She says in a speech to her followers, at Amazon headquarters in Baltimore, "There are misguided women out there who will attempt to remake this world exactly as it once was."[23]

Finally, in his conversation with Rose, a random woman who is also making a pit stop at the monument shrine, Yorick discovers that a lot of women are literally rediscovering themselves in a world where they have become untethered to the patriarchal ordering of their lives. That is, a lot of women are neither helping with the efforts to restore a sense of pre-apocalyptic normalcy, nor are they actively engaged in the process of erasing the last marks of masculinity, but are just doing their own thing, whatever that might be, to make living in the end times as interesting and enjoyable as possible.[24]

Given this state of the union, it is more likely than not that many of the problems created by the gendercide are going largely unaddressed at all or so inadequately as to be mostly ineffective. Which means that even if we allow for the fact that Washington, D.C., was by and large not destroyed by the gendercide, or was the focus of a major post-apocalyptic effort to clean the city and prevent things from getting worse, as Yorick and Agent 355 move farther away from D.C. their world should get increasingly more hostile. This should be the case both in terms of the people they meet and the environment they traverse. And, as expected, this is simply not the case. In fact, the areas that ought to have been the hardest hit by the gendercide are remarkably well preserved and in good working order. Boston, the first stop on Yorick's mission to locate Dr. Mann, is an excellent example of what is fundamentally wrong with how Vaughn and Guerra have chosen to re-present the world two months or so after the collapse of society.

This Isn't the Same World It Was Two Months Ago (Or Is It?)

Boston provides a big window into the presentation of the world because it is the bottleneck of several storylines. First, Yorick and Agent 355 are trekking as clandestinely as possible toward Boston in an attempt to keep secret the fact that Yorick is still alive. Second, a group of Amazons, led by Victoria and guided by Hero, Yorick's sister, is attempting to locate Yorick to finish the job Mother Nature started. Obviously, unencumbered by the need to hide they take a different route from Baltimore to Boston. Finally, a group of Israeli mercenaries hunting for the last man, ostensibly to prevent such a weapon from falling into the wrong hands, is also scouring the countryside for Yorick. Here, it is helpful to pause and consider the likelihood of such a thing.

Vaughn has made clear that the Israeli Defense Force employs women as well as men, mandatorily, so it is completely feasible that a female commando unit of highly trained Israeli mercenaries exists. And, allowing that not every military craft in the arsenal of Israel was airborne at the apocalyptic moment, then it is possible Israeli mercenaries could make the trip from Jerusalem to Boston (or its vicinity) in time to be in the city at the same time as Yorick. And, it is equally plausible that, if they knew with certainty a single male survived, and they were genuinely concerned with what might happen should he fall into "enemy" hands, then it is reasonable that Israel would divert precious resources to a mobile commando unit to secure the last man for Israel.

Driving from D.C. to Boston, however, would normally take about 8 hours, but with the apocalyptic road conditions driving in any sort of direct or quick way would be sporadic and difficult to do—as has been made clear already by the difficulty Agent 355 had in returning the president to D.C. from Arlington. Forced to use a combination of vehicles and walking the trip would be substantially slower going. Knowing how hard it would be to secure gasoline for vehicles, more than two months after survivors would have started siphoning and hoarding so valuable a resource, there would had to have been times when they were substantially delayed in their journey to Boston. But, when next we see the pair, they are camped out in Boston. So, it seems safe to assume that at least a couple of days would have passed between leaving D.C. and arriving in Boston. Of course, being in Boston, even with an idea of where to look for Dr. Mann, the two have been camped out under a bridge for a couple of days as well given that Agent 355 has had time to search for Hero at multiple places, presumably look for Dr. Mann, and Agent 355 is only willing to be out and about during the night—both to help hide Yorick's identity and because she is African American and has trouble "looking white"—so it is plausible that Israeli Mercenaries could be hot on their trail, even if they did have to make a trip across the Atlantic.[25]

When the Amazons, in pursuit, cross the path of a rogue girl in Putnam, Connecticut the scene is leafy green trees, clear night skies, and, again, an area seemingly devoid of debris.[26] All the scenes in Boston are similarly sterile in their appearance—particularly the neighborhood background that passes by as the trio walk down the street. Even when Vaughn and Guerra depict the area around Dr. Mann's laboratory in a wide angle aerial perspective it looks rather like you would expect Cambridge to look as the sun was coming up.[27] Though just to ensure that somehow Dr. Mann's laboratory is not in some special enclave of Boston that was safely inoculated from the apocalypse, consider what Boston should look like to what it actually looks like in the artwork.[28]

Once Dr. Mann agrees to join the duo, the trio is set to strike out for California because the Israeli mercenaries torch Dr. Mann's lab and all of her back up data is in San Francisco. Along with the power outages, and the internet being offline, communication, even in letter form is difficult; so, the only way to access the information and data Dr. Mann requires is to go to it. Leaving Boston is not as easy as getting out of D.C. because there is another person to account for and the trip is not a quick jaunt down the coast, but a transcontinental undertaking. Arranging to leave Boston requires the group to venture into the rail yard. Each of the frames, from Yorick buying passage onboard a departing train, to the train slipping westward out of Boston, to the panoramic view of the train heading into a rural sunset, showcase a Boston still intact.[29] The buildings are not burnt, the windows are unbroken, the streets are clean, and most of the people, though some are obviously frazzled, are contentedly getting by and living a manageable post-apocalyptic life—lives that seem substantially farther along into the apocalypse than the mere months that have supposedly passed.[30]

This Was Fun—in a Perverted Back to the Future *Kinda Way*

Time in *Y: The Last Man* is a bit tricky to deal with because the various story lines must be pulled together simultaneously, over a vast narrative, while conjoining with other story arcs as Yorick moves across the country and, eventually, the world. But more than just trying to keep up with the jumping around, there are some concerns with how time is presented in the story from the beginning. When Yorick and his girlfriend, Beth, are chatting on the phone in the minutes leading up to the apocalypse he is in Brooklyn and she is on a satellite phone in the Australian Outback, but they both appear to be enjoying summertime. Beth, dressed in a bikini top and very short jean shorts, is traipsing around the outback on a sunny day while the scenes around Yorick's apartment depict short sleeves, green leaves on the trees, and equally sunny days. While it could be argued that it just happens to be a sunny day in both places—if the sunrise in Australia is an early six am, then it would be two in the afternoon in Brooklyn—but it could not be summer in both places—and Vaughn commits to summer in Brooklyn when Yorick places the date of the apocalypse as July 17.[31] Moreover, Agent 355 tells the new president that the road conditions are pretty congested because the plague hit at rush hour. Now, it may be that rush hour in Washington, D.C., is a little longer than elsewhere, but one, two, or three in the afternoon—depending on what time the sun rises in Australia—is not rush hour.

This type of time inconsistency is emblematic of the use of time throughout the graphic novel. The obvious rebuttal here is that the story is for entertainment and the action is more important than the consistency of the clock.

However, where the concern is the realistic depiction of the world before, during, and after the apocalyptic event, representation of the world is vitally important for keeping track of how the apocalypse unfolds and is responded to by the survivors. And, to make the point more forcefully, when the pace of the story necessitates skipping huge chunks of time throughout—weeks and months—the time in the story must be accurate or else it creates a layer of distortion that is both an unnecessary distraction and an added element of confusion for accurately determining what apocalyptic conditions Yorick and company should be facing as they traverse the globe.

That said, the story jumps ahead sixteen hours from Yorick's train leaving Boston to his rather eventful arrival in Marysville, Ohio. Marysville is nestled in an out of the way spot within walking distance of the railroad tracks and at first blush the whole area seems to be no worse off than the bigger cities. En route to Marysville, or, actually, on the way to San Francisco, with an unexpected stop in Marysville, Yorick complains that all the little towns along the track look the same because the women have yet to get the lights turned back on.[32] Marysville, however, lights or no, seems to be doing rather well. Agent 355 and Dr. Mann, after exiting the train, are in a field of lush green grass and blue skies, even the power lines are not down, though they clearly are not carrying any power. The people look to be well fed, clothed, and mostly content, as they should be since they live in a town with perfectly normal neighborhoods. Someone who did not know what they were reading, who picked up the book and opened it to the Marysville storyline, would not have any inkling this was a post-apocalyptic story.

Marysville has quite a few idyllic aspects that seem out of place just a few short months after a global apocalypse. But, the experience of Marysville is best explained by those that were there. Dr. Mann describes the medical facility as the "first post-plague medical facility … that doesn't look like it belongs to a fucking medieval barber."[33] Yorick describes Marysville as "a goddamn utopia."[34] And, when the Amazon ladies scouring the earth for Yorick finally catch up to him in Marysville, Victoria tells the residents she has nothing but admiration for them because, "from the looks of [the] community, [they] long ago escaped the yoke of patriarchy [because] no doting wife or blushing bride could have built [such a] kingdom."[35] Finally, when Yorick decides to give himself up rather than see the residents harmed or the community destroyed by the Amazons he explains his decision by saying, "this town is the only decent place I've visited since the world went to shit."[36] Given what Yorick has encountered in the story thus far, from New York, to D.C., to Boston, to Marysville, that last claim is a little hard to believe since most of the hardships he has faced have been self-inflicted in order to hide his identity and nowhere in the novel thus far has been depicted as recognizably terrible.

The first fifth of Yorick's story spans a time period that begins thirty minutes before the apocalypse and ends with he and his trio of companions slinking out of Marysville in the middle of the night a little over three months later. In that amount of time the world should still be reeling from the effects of the apocalypse. Cleaning up three and half billion dead bodies alone would take a lot of time, a concerted effort, and an organized labor force, none of which are present in the story; or, in the case of Yorick's stateside adventure, 162 million dead bodies. Assuming the cleanup went quickly and smoothly, and Waverly was literally picking up the last of the bodies, there would still be a nearly insurmountable amount of damage and infrastructure issues to address. It seems reasonable to argue that there would not be enough women, and not nearly enough skilled and qualified women, to do both tasks simultaneously. Though the experience of the apocalypse would be different in D.C. and Boston and Marysville, the reality is that there is no way all three places could look the way they do, function the way they do, and meet the needs of the survivors in the months immediately following the apocalypse.

The difficulty in tracking Yorick across the country, and subsequently, across the globe, is that book one is the baseline for what to expect in the rest of the story. That is, as time passes in the story there is an expectation that cleanup will have happened in, say Arizona, by the time Yorick gets there, or that California might have been able to return some normalcy to the lives of the survivors by the time Yorick arrives there. The farther Yorick travels and the more time that passes the greater the expectation the reader of the story has that life would have begun to stabilize and return to normal. That doctors would be opening clinics, farmers would be harvesting and distributing foodstuffs, governments would be working to clear roadways and finalize the cleanup and disposal of the dead. The assumption at the outset of the story is that these things would be underway everywhere, but the presentation of the places Yorick has been gives the impression that all of these things have, or have mostly, happened in the two months that Vaughn and Guerra skip in one page turn. The rest of the story's accuracy in portraying the post-apocalyptic world several months or years after the gendercide is undermined by the unrealistic hope cached in the beginning of the tale.

Mother Earth Eliminated Your Kind for a Reason

Book two in the chronicle of the last man begins on a similarly sour note. Only a few pages into the story Yorick is depicted atop a mountainside cliff, waiting for their train out of Marysville to undergo some quick repairs, and the scene is as idyllic a panoramic vision of middle America that can be conjured.[37] The scene in Kansas, when the group arrives ten hours later, is equally quaint and picturesque. In the first 40 pages of the second book, there is a mention of failing cell towers and one nuclear power plant failure in Eastern Europe. However, both of these instances are treated like isolated incidents in a grander scheme of stability. As if, in the post-apocalyptic world of the gendercide, disasters are still happening the way they did pre-apocalypse. That is, a major disaster like a nuclear meltdown is manageable even in the post-apocalypse and this perception is largely because it is understood outside of an apocalyptic frame of reference. Though Natalya, a Russian government agent on a mission to rescue a cosmonaut, claims there are possibly one million women presumed dead because of the nuclear meltdown in Eastern Europe, Agent 355 assures Dr. Mann that nothing similar could happen in the United States because her fellow surviving secret agents have been decommissioning nuclear facilities for months.[38]

Even allowing that Agent 355 is telling the truth and that her counterparts are capable of preventing every nuclear facility in the country from becoming a post-apocalyptic Chernobyl or Fukushima, there are 450 active nuclear power plants located in 30 countries and 60 more are under construction in 15 countries.[39] Given the incredible difficulty of traveling throughout the United States it is unlikely the secret agents would be able to make quick work of shutting the facilities down. Even if decommissioning nuclear facilities was priority number one for the remaining women in government there are potentially 45 other countries with active nuclear facilities that are not under the purview of Washington, D.C. Even were they to be moving from one country to another, bypassing those governments or assisting them in managing an emerging nuclear threat, it would be a tremendous undertaking in the immediate aftermath of the apocalypse to travel internationally. Bluntly, in spite of Agent 355's optimism and assurances, there are definitely nuclear disasters occurring globally.

Many of the active reactors today were built between 1970 and the mid–1990s putting many of them at risk for failure now, but especially so in the aftermath of a cataclysmic event. Fukushima was damaged by an earthquake and subsequent tsunami and though reports that it has gone on to irradiate the Northern Pacific are largely exaggerated, it is nevertheless a concern. That the ocean is large enough and deep enough to absorb the radioactive fallout of one meltdown that alone does not indicate that multiple meltdowns could not wreak havoc on ocean ecology and terrestrial ecosystems could easily be wrecked as a byproduct of distant nuclear failures. In an apocalyptic scenario, not only would wind, rain, and water move the radiation around, but disposing of irradiated bodies would be more difficult not to mention the site cleanup if it was even possible. It is hard to imagine the apocalypse as a hardship if all of the potential hazards are easily and quickly rendered moot. Of course, Agent 355's reassurance is predicated on a given notion that all of the reactors and nuclear facilities in the United States—and elsewhere—

managed to escape any and all tertiary damage at the moment of the apocalypse. Of course, if all the planes fell harmlessly into fields, maybe all the nuclear facilities did escape unscathed; the Kansas storyline presents a view of the Midwest after the gendercide that strangely reinforces this notion.

Kansas is presented from two viewpoints in the book: one, Yorick and his companions traveling via train, truck, and on foot as they move through the landscape and, two, the Israeli troops tracking Yorick who, airborne in a stolen helicopter, offer glimpses of the landscape from above. Neither group encounters anything in their backgrounds that would suggest the residents of Kansas have not managed to quickly claw their way back to normalcy. This, of course, should come as no surprise given that their arrival in Kansas is one year after the gendercide occurred.[40] After a year there is an expectation that people would be getting past the horror of the aftermath, but as I discussed above, this state of affairs seems to have been the norm from the beginning of the post-apocalyptic era in *The Last Man*.

Inexplicably the group heads for Nebraska after they leave Kansas and the scene in Nebraska is just like Kansas.[41] Unsurprisingly, when the group make their way to Colorado there is, again, nothing to suggest that the area survived an apocalypse because there is nothing to suggest that an apocalypse occurred.[42] Because the trio is taking the road less traveled there is, perhaps, a greater expectation that the setting for their Colorado adventure would produce some indication that an apocalypse occurred. The group is "about a two-hour walk" to civilization from their mountainous path, but there is, rather than an apocalyptic indication of any sort, a perfectly tranquil scene with green trees, a burbling stream, and an undamaged bridge in the background.[43]

Finally, the trio ends up walking down an interstate in Arizona eighteen months after the gendercide occurred. Dr. Mann remarks that given the amount of time that has passed since the apocalypse began it is likely that the pygmy shrew has gone extinct.[44] Soon to be followed by possums and rats, species that have short life cycles and have been, like all the other species, robbed of male counterparts. The Arizona leg of their journey is less important for its continued portrayal of a healthy landscape, and not even really for the aside of marking the passing into oblivion of several species—though here it is important to note that Dr. Mann's concern is for a species she thinks is the first to go extinct because it could not mate, ignoring the possibility that other species have already been destroyed by other post-apocalyptic circumstances, i.e., human activity.

No, the Arizona segment of the story stands out because it contains a rare encounter with the apocalypse in the form of a deceased all-male road crew, now bones in coveralls, piled up in front of a derelict bulldozer. Agent

355, standing solemnly at the edge of the grisly scene, remarks that it is "another road crew," which suggests this has been a frequent occurrence, or, at least, frequent in Arizona. Interestingly, when Yorick lambasts in absentia the many women who must have driven by those bodies, which he is certain has happened frequently, because the "roads are open [in Arizona because] it wasn't rush hour when the plague hit" the western time zone, Agent 355 responds, "You know the story. When three billion people die in a day."[45] Farcically, the moment is ruined when it is revealed that the bodies have been staged there as a trap by a rogue element of anti-government women.[46] The way Vaughn addresses certain apocalyptic concerns in the dialogue but does not match those concerns to a representation of the world betokens the exact problem narratives of this kind pose for a real conceptualization of the fragility of the environment and human society faced with a large scale, global catastrophe.

After the Arizona sequence, the story springs forward to California where, having started in Boston, as an attempt to furnish Dr. Mann with the specimen necessary to develop her cloning research and find a cure, forced instead to seek out her alternative lab in San Francisco, the group finally arrives nearly *two years* after the apocalyptic event. Two years seems like a reasonable amount of time to expect that the post-apocalyptic world, having not devolved completely into anarchy and destruction, would start to bounce back in important ways despite the looming threat of total extinction facing the remaining inhabitants of the earth. But, in San Francisco, civilization has progressed well beyond that, at least at first blush, as Yorick attends an organized professional level basketball game disguised as the team mascot for his birthday.[47]

Additionally, the streets are incredibly clean and the lights around town are back on—assuming they went out at all, but since the East Coast suffered a period of blackouts, energy inconsistencies, food shortages, and other problems, there is no reason to assume the West Coast faired any better. So, having lights, clean streets, professional basketball, operating trollies, and an organized, working police force, and other luxuries of the pre-apocalypse it seems more than reasonable to suggest that California is recovering nicely. Of course, it is not just San Francisco that appears to be doing well. If the California story arc only focused on San Francisco it would be reasonable to argue that San Francisco was spared destruction or was unusually well suited to recover quickly from the plague's aftermath. And, maybe, all of California was just as lucky, because the trio heads to San Diego to jump aboard a ship, and San Diego is presented as beautifully as San Francisco.[48] At this point in the story, after making a transcontinental crossing over a less than apocalyptic country, Yorick's adventure as the last man goes global.

I Wish I Could Say This Was All Just a Drill

Since the point at which Yorick's story goes global is, chronologically, so far removed from the apocalyptic event there is a reasonably high expectation that the world will be put back in order though, obviously, there are some aspects of the post-apocalyptic that ought to defy these expectations. Primarily, it seems as though many of these expectations for the world would center on social organization, stability, and the delivery of necessary goods and services to the remaining citizenry. How this is achieved uniformly across disparate pockets of population that are geospatially disconnected is a difficult aspect of the post-apocalyptic to work out. Bearing in mind that many women in the story have already broken free of a conception of the world tethered to the pre-apocalyptic ideal and the Amazonian tribes—which, in spite of a lack of communications and restricted travel have a global reach—have established goals that are far removed from those of the provisional government in D.C. As a result, the governments at the state and federal level must be working in tandem *for* particular ends while working *against* adversarial components attempting to thwart their efforts.

This is also being done, recall, in a social milieu that includes women assuming alternative hedonistic lifestyles that fully embrace the "end is near" narrative that must necessarily accompany a realization that the human species is nearing total extinction. Regardless of who is working to regain social, economic, and political stability, even if doing so meant wide scale and successfully coordinated campaigns to address the severest aspects of the apocalyptic fallout, there would be issues that would remain unaddressed for a long time as the ability to deal with such issues would be beyond the reach of the survivors to contend with and even years afterward the evidence of the apocalypse should be visible to some degree.

In this context Yorick and his companions make their way toward Japan aboard a repurposed cruise liner. Vaughn accurately captures the reality of Captain Kilina's ability to be at sea as one of the very few women with no military training that possesses the ability to pilot a ship, especially one in the hundred-ton class. But the boat and its quasi-piratical crew, along with an adventure on the high seas, detracts from a reality that is not only incorrectly backgrounded but scrubbed clean. The oceans are currently a mess and in a post-apocalyptic world where shipwrecks would become common for a long time after the gendercide, their effects would linger for considerably longer since oceanic cleanup would not be high on anyone's priority list, or at least not on any list that would enable people to make a significant difference. Currently, "there are 5.25 trillion pieces of plastic debris in the ocean. Of that mass, 269,000 tons float on the surface, while some four billion plastic microfibers per square kilometer litter the deep sea" according to scientific estimates.[49] This is just debris that exists now.

Granting that most of the trash that accumulates in the ocean is plastic, and much of that plastic is microplastic bits, the worst area in the world for the accumulation of the garbage in the ocean is the North Pacific Gyre. The area through which Captain Kilina is sailing her vessel. However, the situation at sea would be far worse than it is currently because there would be nothing acting as a stop gap measure to prevent a lot of additional garbage from making its way to the ocean during the two years it takes Yorick to get from D.C. to Kilina's ship. The most recent scientific research suggests that humans collectively put an estimated "8 million metric tons of plastic trash" in the sea every year which amounts to "the equivalent of five plastic bags filled with trash for every foot of coastline around the world."[50] If society suffered a massive breakdown resulting from an apocalyptic event, then not only would more trash and debris find its way to the sea, but it would join the debris created from shipwrecks.

Some of the ocean liners that would crash would be carrying oil and other hazardous cargo. Oil slicks alone would not only kill thousands of aquatic species, but the slicks would be present for longer than usual. The Deepwater Horizon oil spill in the Gulf of Mexico took more than three years to clean up and efforts to clean up the spill were abandoned before they were fully complete to the satisfaction of non-industry evaluators. Once the spill area was returned to pre-spill levels most cleanup efforts were called off. However, before the cleanup could be considered officially over BP reported that it had expended more than $14 billion, required more than 48,000 employees, and totaled more than 70 million personnel hours to accomplish its *still incomplete* cleanup goals.[51] This level of concentration of money, resources, and labor would not be feasible in a post-apocalyptic world, but even if it was it would be difficult to address the numerous spills individually, but taken collectively the oceans would be literally wrecked.

In 2010 the world's maritime fleet boasted approximately 4,000 tankers hauling oil, gasoline, and other petroleum products.[52] However, the amount of oil shipped by sea has multiplied twenty-seven times since 1935 and the global maritime fleet employed roughly 11,000 tankers by 2013.[53] Oil shipments account for roughly 30 percent of maritime trade and hauled an estimated 1.78 billion tons in 2012, while shipments of refined products and liquefied gas amounted to 1.05 billion tons (of which approximately 230 million tons were liquid natural gas, or, LNG) according to the United Nations Conference on Trade and Development (UNCTAD).[54] Assuming just one-third of the global fleet was at sea and did not have a female crew member on board that was capable of successfully piloting the tankers, then there could easily be hundreds of millions of gallons of oil, gas, and other chemicals at sea which, in addition to dirtying the water and coastlines, would coat the garbage, kill oceanic life, and, potentially, be on fire as well.

Compare this harrowing minimalist scenario to the representation of the oceans in Vaughn's story. When Yorick, Agent 355, and Dr. Mann arrive in San Diego the docks are not only clean but so is the water.[55] Moreover, all the visual representations of the sea are impeccably clean in appearance to the point they are not even realistic representations of the ocean in a pre-apocalyptic sense. The initial depiction of the sea is a night scene and might provide adequate cover for the absence of any detritus or apocalyptic debris.[56] Subsequent portrayals in the initial ocean voyage scenes are also at night, and in relatively calm waters, including a full page shot of a fully submerged Australian submarine patrolling the waters.[57] Unfortunately, when the story shifts from the cover of darkness to the next day there is nothing substantially different in the way the ocean is portrayed—for either the frames showcasing Kilina's ship or the submarine in pursuit of them. The failure of the apocalyptic world to show up in the ocean scenes continues through the depth-charge attacks on the sub and the torpedoing of Kilina's boat. Not until the submarine emerges from the ocean after the sea battle is any debris visible in the ocean and this is from the wreckage of the torpedoed ship.[58]

When the survivors of the battle finally make it to Australia the country looks well organized and operating with some degree of normalcy though there are telltale signs of the apocalypse and, perhaps, evidence of the difficulty of recovering from the gendercide. There are buildings still in rubble, broken windows, and inflation adding to the troubles faced by Australia's female residents, but the country has a telegraph up and running, a functional airport, and a dock that can accommodate trade. Though Australia appears to have had a harder time dealing with the gendercide, especially by comparison to the depiction of the United States, the country does come across as stable and rebounding given the circumstances.

The stateside world is kept in the picture by Hero, Yorick's sister, who is making her way from California back to D.C. She departs California around the same time Yorick heads to San Diego and travels roughly the same route back to D.C. that Yorick and his companions took to California. This sequence allows the reader to revisit many of the same places and the message is clearly that America is once again thriving.[59] Similarly, when the group makes it to Japan there is no longer a sense of apocalypse. The first scenes in Japan reveal clean oceans at the docks, a bustling community scene with cultural theater as the centerpiece of the sequence.[60] More than that, when the foursome splits up in Japan neither twosome experiences anything but pristine countryside whether on foot or by train.[61] In fact, having visited Australia, Japan, and Papua New Guinea, and including Hero's journey through the American heartland, the only real insinuation that things are still going poorly outside of an acceptable norm is a concession that the president needs to do something about the "borderlands" which are becoming "a nightmare."[62] What exactly constitutes the nightmarish conditions of the borderlands, and how extensive the borderlands are, remains unexplained, but clearly the areas of concern at this point in the story are marginal.

No Offense, but It's Not All About You

The final leg of Yorick's story, book five, provides a view of life in China that is not substantially different than that of Japan, but it is in China that the foursome becomes a twosome as Yorick and Agent 355 head for France while Rose and Dr. Mann, who now has what she needs to complete her research and develop a curative to the gendercide, remain in China at a cutting edge laboratory that will facilitate Dr. Mann's research. The story continues to oscillate between Yorick's adventure and Hero's own journey as both are now making their way to France for the culmination of the story. Early in the final leg of the narrative the Israeli troops pursuing Yorick by tracking his sister and her companions are shown sailing through the untroubled waters of the North Atlantic.

Like the Pacific, the Atlantic appears to have escaped the apocalypse unscathed, and the narrative of the apocalypse is all but over except for the persistent need to find a solution to the gendercide. Most of the locales in the global scope of the story are picturesque,[63] and when they are shown to have suffered from the chaos of the apocalypse, they still have not devolved into a state of nature situation. Trains and rail systems are intact and running around the world,[64] mail is being delivered again,[65] high level scientific research is taking place, governing and commerce are on-going, and though some of these social aspects would be expected five years after the apocalypse, some societal functions would be, if not permanently, substantially undermined even at this point in the story.

Once the tale of the last man on earth reaches its penultimate action scene Vaughn skips ahead sixty years. Not only has Dr. Mann managed to utilize Yorick's and Ampersand's DNA to code against the gendercide plague, but she has also successfully created clones. The scene that is Paris, France sixty-plus years after the gendercide is one of magnificence and opulence that easily surpasses the pre-apocalyptic world.[66] The world has become futuristic in the most relevant sci-fi ways betokening a future that would only have been possible for a world that had to claw its way out of an apocalyptic nightmare. Here, more than at any other time in the story, Vaughn and Guerra create that dystopian hope that reinforces not only that humanity could survive a near extinction event, but that the long-term recovery would, presumably because the survivors are working with a largely blank slate, produce a world that was substantially *better* than the pre-apocalyptic world.

In a series of flashbacks the intervening years are depicted from Yorick's point of view as he travels the globe visiting his family and friends. His sister and former girlfriend are shepherding the African wildlife back to health by protecting and monitoring their species well-being. Research is on-going at Dr. Mann's lab producing males of every species types to repopulate the earth, and an aged Yorick makes his way through a wintry landscape to bury his longtime pet. All over the world things have not merely returned to a pre-apocalyptic normal, they have not even returned to a recognizable version of pre-apocalyptic normal, but rather have surpassed the sense of normalcy which seemed to plague the survivors in the years that Yorick, Agent 355, and Dr. Mann were trekking across the globe in search of a cure and, what is more, there is no inkling whatsoever that this is not *abnormal*.

Like *The Walking Dead* the story of Yorick as the last man on earth begins with a viral plague capable of destroying enough of the human species quickly enough that all social order is also destroyed. However, both of the plague stories are focused on the deaths of billions of people and the plight of the survivors, but neither deal with an important aspect of viruses in the Nuclear Age. Though Kirkman's tale resurrects the dead and Vaughn's story uses a manmade genetic disease, neither of these stories deal substantially with what might happen to the survivors of a plague that was itself a mutation. In Jeff Lemire's *Sweet Tooth* mutations play an important role and in *Snowpiercer* the fate of the survivors is entangled with a nuclear vision of the apocalypse that afflicts the survivors. So, in turning away from the end of the world as a struggle to overcome death, our focus will now shift to the end of the world and the struggle to live in a future determined by the apocalypse not just by the social forces of a human response.

5

Post-Human Life in a Post-Nuclear Age in *Snowpiercer* and *Sweet Tooth*

Fantasias of Possibility

Jeff Lemire's *Sweet Tooth*[1] is not a nuclear story in the same sense that *Snowpiercer* is, but Lemire's story does share some story elements that make it a nice bridge from the in-depth study of viral apocalypses in the previous chapters to a more realistic form of apocalypse represented by the nuclear age. In *Sweet Tooth,* Lemire creates a world where a nuclear age scientific endeavor unleashes a plague on the planet that slowly eradicates humankind. Though his science-gone-bad approach is tethered to some otherworldly possibilities, the laboratory aspects of his story certainly coincide with the plague narratives in the previous chapters. In fact, high-tech cataclysmic narratives have a pedigree that predates the first weaponized use of atomic science by almost 65 years.

Beginning with heavier-than-air flight and the frightening possibilities that posed to the fin-de-siècle imagination when combined with industrialization catastrophic questions emerged across a broad range of potential threats. "Superweapon narratives featuring radioactive waves, electrobombs, weaponized bacilli, sterilizing rays, and even atom bombs flourished in the early years of the twentieth century."[2] In their air war context these futuristic musings about airplanes, bombs, poison gas, the metropolis, and a military-industrial complex that operated the socio-technological aspects of society are uncannily reminiscent of our current attempts to conceptualize the *possibility-as-cataclysm* we are ecologically confronted with today.

Lemire's story centers on Gus, a human-animal hybrid, who, unbeknownst to himself, holds the key to deciphering the apocalypse. When Dr. Singh, a doctor searching for a cure to the plague, locates the facility that spawned Gus, he finds an underground, top-secret lab where genetic cloning was being done on ancient skeletons.[3] Using testimonial evidence he has collected from Gus, and hard evidence he has gathered about Gus's life, Dr. Singh pieces together the reality of what the government was doing in the Alaskan bunker laboratory and, whether the experiments were playing god, as Singh claims, or just overzealous scientists whose experiments got away from them is largely a mystery. The reality for the world of *Sweet Tooth* is that the virulent nature of the plague cannot be stopped, and humans are ultimately doomed.

The concept is not that far removed from either of the approaches employed by Kirkman or Vaughn to build their apocalyptic narratives. However, unlike both *The Walking Dead* and *The Last Man*, Lemire's story, though his plague has early effects on human society eerily similar to the viruses in Kirkman's and Vaughn's stories, picks up a decade after the plague has struck and presents a world that is obviously more stable than one would expect given what we know about the problems facing the world if a viral apocalypse wiped out a majority of the human race. Though much of the human built world is presented as dilapidated, dirty, broken, and run-down it is important to realize that these structures are still serviceable in many ways *despite being mostly abandoned for the better part of a decade*. Early on in the story, when Gus is forced to leave the relative security of his nature preserve home, Jepperd, the good-hearted rogue that finds Gus, says "the sick hit seven years ago" around the time that the hybrids started to show up.[4] But, when the duo finally start to make their way through the wider world of Nebraska the phone poles and utility lines are still standing or, for the most part, leaning, and many of the structures the two pass by or seek shelter in are in amazingly good shape considering the amount of time they have gone without being maintained. Additionally, the landscape is incredibly clean, free of debris, and hardly overgrown at all.[5]

Lemire's story is guilty of the same sorts of environmental omissions that vex Kirkman and Vaughn, but because Lemire's narrative utilizes a much slower death of the total population it is much more reasonable to assume that some people were able to control for many of the factors that would have seriously complicated life in the post-apocalyptic world of *Sweet Tooth* had Lemire's virus been as effective as Kirkman's or Vaughn's. More importantly, and more fairly, because the virus in Lemire's story does not kill immediately or rapidly work its way through the population, it is possible that some individuals, once they became sick, made every effort to attend to important needs like properly shutting down facilities, or removing hazardous wastes from the roadways, etc.[6] Still, the human characters in Lemire's story face the hardships of dwindling resources, continual death from the plague, and social instability as governments fail and regional disputes erupt across the country—and world, but the whole of the story that Lemire tells is situated

5. Post-Human Life in a Post-Nuclear Age 103

in a geographically narrow strip of land, mostly in the continental United States, encompassing a swath of area roughly one thousand miles long from Minnesota (briefly) to Colorado as well as an isolated military research facility in Alaska.

The aspects of Lemire's story that link to the viral tales of Kirkman and Vaughn are, or ought to be, embroiled in the same sorts of problems, but since the apocalyptic world is only shown in flashbacks it is difficult to get a sense of what the world "looked like" rather than just the experiences of the characters as they reflect on their lives in and immediately after the apocalypse. This is especially pertinent to crafting a robust notion of the post-apocalyptic world that Gus grows up in because, even if there was minimal environmental fallout, there is an exponentially diminishing possibility that the survivors clean up the world, and, after a scant ten years, the world would not have been able to fully digest several centuries of industrial life. Yet, whenever Jepperd and Gus traverse a landscape that is presented as an expansive, panoramic vista, the post-apocalyptic world appears unsettlingly pastoral and pristine—except for those areas that are urban or occupied by survivors that have gone native.

Lemire shows the pre-apocalyptic world primarily as it is in the process of falling apart as plague panic sweeps across the country, but rarely does he depict the immediate aftermath of apocalypse at length. Rioting, the destruction of urban environments in the panic of people frantically searching for safety, and the breakdown of government control, are all present and on display, but the environment is not represented on a large scale. The post-apocalyptic world, contrarily, is on full display but, in terms of time, spans only a few months, at best, from the start of the action to the conclusion of the story proper—the last issue of the saga is dedicated to a longer generational view of the remaining characters. The difficulty, then, for an apocalyptic analysis such as this is that it is entirely possible that the rest of the world is a wreck and in no way comparable to the area Lemire makes the setting of *Sweet Tooth*.

It is tempting to surmise that other places are far worse off than the Front Range of the Rocky Mountains and the frozen heart of Alaska, but there is no reason to assume that. Instead, the dystopian reality that Gus and his fellow travelers face—both his companions and nemeses—is the end of the apocalypse and the beginning of a new age of peace and prosperity. The apocalypse only lasts roughly ten years before the post-human era dawns. The plague is terrible for pure humans start to finish *as a disease* because there is no cure, it is obviously an awful way to die, and eventually all humans will succumb to it. But the apocalypse is not really that hard on many of them as they are able to maintain some semblance of social order, though on a much smaller scale, and they are able to successfully scavenge plenty of resources to sustain themselves fairly well even a decade into the apocalypse. Not only is there fuel available but often stores have not been completely looted allowing the characters to find many items needed for survival, in a state of good repair, with little effort.

Before You Can Become Something New, You Must Remember Who You Have Been

The remaining characters at the end of *Sweet Tooth*, especially the hybrids, are the important point of departure from the other plague narratives and an important connection to the nuclear world of *Snowpiercer*. For Kirkman's survivors what caused the virus and why some people were immune to it, at least partially, remains a mystery and, similarly, for Vaughn, Yorick is immune only because he has a pet monkey that was genetically manipulated to be a carrier of the plague and as such bears the necessary immunity too, which Yorick was exposed to as his handler. In Lemire's story the animal-human hybrids that are born as a result of the plague are naturally immune to the disease and, hence, come to represent the best hope for the continuation of the human species in the post-apocalyptic world because every human remains susceptible to the plague even decades after it first strikes.

Sweet Tooth provides a post-human look at an apocalypse, that is, what life would be like in a world where the apocalypse required an evolutionary shift in the physiological and genetic make-up of the human species. *Sweet Tooth* follows the journey of Gus as he ventures for the first time beyond the boundary of his nature preserve hideout to his ascension as the undisputed leader of the hybrid species as they establish themselves in the world as the dominant species, supplanting the dying out non-hybrid humans. Though the hybrids are animal-human mash-ups they also have a deeper connection to the earth and nature representing an aspect of early human society that was, according to Gus, lost as humans evolved socially.[7] It is increasingly a familiar trope in apocalyptic narratives that the survival of humans and human society hinge on the emergence of a trans-human or post-human development. The apocalyptic not only erases the socio-cultural status quo allowing for a better future but provides the springboard for the development of better humans to inhabit that future.

That the idea of the post-human should find a home in the apocalyptic is unsurprising given the various discourses of contemporary science and science-fiction. The belief that a better future for humans will require better humans is increasingly potent and, harkening back to Rick Grimes at the end of *The Walking Dead*, even he is concerned with being better than everyone's expectations of him so that he will be fit to lead them into the better future. The idea of the better human, the superior existence that is meant to inherit the better future, has its roots in the rich histories of several religions so it is unsurprising that as dysto-apocalyptic narratives gain in popularity, their central characters should become salvific in both spiritually and physiologically transcendent ways. Though it does not begin that way, the positive outlook of the *Snowpiercer* epic hinges on a very similar notion about the necessity of transforming the human element into something that can live in a post-apocalyptic world.

Know Your Place, Keep Your Place

Snowpiercer[8] is a tricky saga to trace because it was written over a period of time stretching from 1984 to 2016, it was penned by different authors, who attempted to keep the internal narrative consistent while also being responsive to the real-world conditions in which a particular contribution to the story was being written. The result, unfortunately, is a story that has significant internal inconsistencies which make it difficult to represent the world of *Snowpiercer* coherently. As such it is important to take each segment of the story individually first, and then examine the larger issues presented by the overarching narrative. Adding to the difficulty of using *Snowpiercer* for the purposes of this examination of apocalyptic ecology is the fact that the world is snow and ice bound rendering the entire planet an easily describable and decipherable inhospitable frozen waste. However, the structure of the narrative does fit the analysis so far in part *because of the disjointed nature of the story*. That is, the hopelessness that Jacques Lob creates in his microcosm aboard the train Snowpiercer is combatted and balanced by the hopefulness that Olivier Bocquet introduces into the world of *Snowpiercer*.

The original *Snowpiercer*,[9] a Cold War story set in the aftermath of a nuclear winter, was published in 1984, the same year, coincidentally, that Carl Sagan published his now notable book on the aftermath of nuclear war, *The Cold and the Dark: The World After Nuclear War*.[10] *Snowpiercer* takes its name from the train that houses what remains of humanity in the persistent global winter that follows the fallout. The environment is presented as a frozen wasteland, one that encompasses the entirety of the world, leaving all known survivors, the inhabitants of the train, stranded in perpetual motion following the same path around the globe. The first description of Snowpiercer given by Lob is that it is a train one thousand and one carriages long that never stops traveling "across the white immensity of an eternal winter."[11]

As it is presented in Lob's version of *Snowpiercer*, life outside the train is not possible given the extreme wintry conditions and freezing temperatures, hovering somewhere around minus one hundred and twenty degrees. Indeed, at the outset, when Proloff, Lob's main protagonist, makes his escape from the impoverished tail of the train his survival is remarked upon as being incredibly lucky because "many others have died in the attempt" even heavily wrapped up as he was it rarely provides protection against the cold.[12] As a result of the bleak and frigid environment outside the train post-apocalyptic life on the train quickly takes up precedent as the most important facet of the story. Proloff highlights the conditions of life in the rear of the train saying that he would rather risk "the white death" over continuing to live in the tail's dreadful conditions.[13] Snowpiercer, similarly, is always shown with its cowcatcher blasting through unending snowy obstacles that blanket the tracks in an "endless waste" and, subsequently, the setting for *The Escape* is always rendered rather bleakly in its black and white hues.[14] The depiction of the world captures, quite nicely, the eerie implication of Derrida's work, as related by Paul Saint-Amour, who claims, "by negating the possibility of a symbolic aftermath, the nuclear condition afflicts humanity with a case of anticipatory mourning, a mourning in advance of loss because the loss to come would nullify the possibility of the trace [of a symbolic order]."[15]

Even as the story presents a dire warning about life in a post-apocalyptic world destroyed by nuclear war, the lesson is not necessarily one of anti-nuclear development, but rather a socialist critique of capitalism and classism. Proloff's first evening outside the confines of the tail finds him drinking coffee and eating train grown vegetables for supper served with fresh baby mouse, purposefully grown for food, unlike the rats Proloff is used to incorporating into his meals.[16] The stark contrast between the lives on the train is brought into sharp relief during this meal as the social divide between the tail and the rest of the train is discussed. In one of the few depictions of the apocalyptic event Proloff recalls the boarding of the train—an event he grimly describes as a massacre—as thousands of people were attempting to force their way onto the train.[17]

He further recalls the tail of the train being added in the immediacy of the moment, not as something done with forethought, betokening a sense of the world that certain lives were expendable. In fact, the reality of that classist worldview is on full display later when the first-class leaders reveal that they intend to disconnect the tail of the train in order to make their own lives more comfortable. Interestingly, the way that Proloff explains the start of the apocalypse, getting on the train was obviously widely recognized as the only way to survive what was coming even though, well before the apocalypse occurred, the oddly named Snowpiercer was built and advertised as a pleasure train.

Adeline Belleau, a second class pro-tail activist sharing Proloff's meal with him and trying to learn about the living conditions of the third-class, recalls that "the screams [of those trying to get on the train were] strangely muted in the wintry air, as if petrified by the cold…"[18] If it were winter when the apocalypse began, then the panic would have been understandable as the world would have already been in hibernation and any survival would have been more difficult. That is, the world Proloff and Adeline were boarding the train in would already have been frozen, covered in snow and ice, with little to glean from the landscape by way of survival for those that did not know of or could not make it onto the train. Surviving in such conditions would have been harder still because the cause of a nuclear winter is not the warheads themselves but the amount of ash, smoke, and dust thrown into the atmosphere from both the detonations and the contributions from the firestorms that would result from each nuclear explosion.[19]

However, though it was snowing in Adeline's memory, Proloff reveals later that the cataclysm began in July which makes for a more believable scenario where panic would have been the norm—after all, how often do snowstorms happen in July in the Northern Hemisphere? The focus on a nuclear winter sparked by a nuclear device stems from Proloff's assertion that a bomb destroyed the climate.[20] Further, he argues that if the cataclysm's cause was something other than the bomb, then it was strangely timed to start at the same time as the war.[21] The combination of these two claims, along with Proloff's description of "a strange wind" and a "terrifying blast that swept everything away … life … civilization … in just a few hours,"[22] is the basis for an interpretation of a nuclear winter as the result of nuclear war; but, in all reality, that need not be the case, as Proloff's argument is balanced by the claim that they really do not know what caused it.[23]

The Infinite White

Generally, to the extent that such as event is possible, there is a consensus that it would take roughly 100 nuclear explosions to produce a nuclear winter—the actual number being a variable dependent upon how large and how technologically advanced the detonated nuclear devices are. Additionally, determining these variables would establish the severity of the nuclear winter that would follow. The actual mechanism driving the winter event would be the resulting firestorms rather than the detonations themselves. Each detonation would contribute significantly to the ferocity of the aftermath, but it would, again, be a highly variable question regarding the extent to which the detonations would contribute. Each explosion would throw up a lot of ash, dust, and smoke individually, but it would take a lot of cloud cover to blot out the sun long enough to generate long-term winter conditions. Hence, the resulting firestorms, rather than the explosions, would be the determining factor for the severity of the winter because it would matter how much combustible material was in the proximity of the blasts, how and whether the fires could be contained or snuffed out by weather conditions, as well as how the smoke, ash, and dust was atmospherically collected and circulated in the air.

But, consider the size of the firestorm that resulted from the atomic bombing of Hiroshima, an estimated 4.5 square miles, versus the one that resulted from the non-nuclear bombing of Dresden, an estimated 8 square miles.[24] None of the bombs deployed in the firebombing of Dresden were nuclear, there was just a massive quantity dropped, and the post-bombing destruction still managed to outpace the nuclear destruction of Hiroshima. Alternatively, bombs, nuclear or otherwise, are not required to produce the conflagrations that could create nuclear winter conditions. For instance, a geothermal super-eruption, like the Old Faithful geyser at Yellowstone National Park exploding from deep with the earth, could also cause similar conditions, especially if there were two or more simultaneous explosions—explosions that would dwarf the destruction of Krakatoa.

A supervolcano eruption at Yellowstone would affect a circular area roughly 500 miles (800 km) across and could affect geography, watersheds, and climate as well as disrupting agriculture in the Midwest and Western United States.[25] Though the lava flows of such an event would likely only threaten an area the size of the park itself, the gas and ash ejecta could alter the climate for upwards of a decade in the areas most directly affected by the explosion. The expulsion of sulfur dioxide would form "a sulfur aerosol that [would absorb] sunlight and [reflect] some of it back to space. The resulting climate cooling could last up to a decade. The temporary climate shift could alter rainfall patterns, and, along with severe frosts, cause widespread crop losses and famine."[26]

Although scientists are ready to acknowledge that super eruptions are a real, if improbable, possibility, though a small one, there has not been one in more than half a million years. Scientists are also keen to point out that in the event of a supervolcano erupting, though it would be devastating, life would continue and much of the world would be unaffected. However, in an apocalyptic vision, on par with *Snowpiercer*, it is possible that if all 28 known supervolcanoes—or even just half of them—erupted at the same time the result would be incredibly catastrophic for life on earth. Such possibilities certainly underlie the idea that the cataclysms origin is uncertain as well as the reality that the apocalyptic conditions of *Snowpiercer* need not be the product of nuclear war.

Endless Frost and Desolation

Assuming that Proloff is correct in his summation of the cause of the perpetual winter, *Snowpiercer*, by presenting the world as a stark and frozen wasteland, cheats its readers out of an opportunity to understand the real costs of a long-term nuclear winter. Regardless of whether one is examining life in a nuclear winter, a supervolcano holocaust, or the freezing conditions of a global warming triggered ice age, the social critique is just as important as understanding what would happen to the environment besides being iced over. What is waiting outside of the train is a better context for motivating the various social concerns aboard the train. Each time the train is shown from outside it is accompanied by a description of the world as desolate, empty, utterly destroyed and inaccessible, yet, though the wintry conditions have persisted for years the train occasionally rambles through an urban landscape that is still discernable and incredibly unburied.

At one such moment, Snowpiercer travels through a rail yard and if it were not for the already established reality of a never-ending winter it would be hard to differentiate between the nuclear winter conditions and a run-of-the-mill harsh winter. Snowpiercer makes its way past long abandoned railcars that have not been moved in years while snow has, presumably, regularly fallen, and yet their wheels are not even covered. Nor are the rooftops of the buildings that breakup the background buried in snow, their irregularly shaped outlines clear, despite years and years of snowfall, ice, and heavy winds. To complete this disjointed appearance of the world the ethereal narrator of Lob's work intones, "without a destination, the train travels on, through endless frost and desolation. Outside, life has vanished from the frozen earth. No promised land awaits these weary, eternal travelers. The promised land is lost."[27]

In fact, as bad as things are on Snowpiercer they would be far worse off of it. Even if the freezing conditions subsided enough to allow for life to resume off the rails, doing so would be nearly impossible. The effects of the nuclear winter would leave behind widespread famine conditions along with toxic radiation levels at every detonation site. Most projections for the length of a nuclear winter—dependent on several highly variable factors such as how many and what types of bombs are used as well as how widespread the post-detonation destruction is, including how broadly deployed the bombs and devices were—suggest that Snowpiercer would be traveling on its course for several decades. Given the timeline Proloff and others hint at it seems as though Snowpiercer has been on its journey for almost two decades, but to be fair to the apocalyptic idea driving the story, it seems like the passengers ought to expect to be onboard the train for about two generations, roughly five decades. In the fifty years or so that the inhabitants would need to be onboard the train the world would be suffering from a persistent nuclear winter during which everything outside the train would die.

First, the snow would make foraging difficult for any person or animal that did not make it aboard the train, but eventually photosynthesis would stop leaving herbivores with nothing to eat. Not long afterward omnivores and carnivores would exhaust their food sources as species propagation would slow down until it stopped as well and, shortly after that, scavengers and eaters of carrion would cease to exist as well. Even if it took a full decade for all non-human animal life to die off from slow starvation and radiation, the humans aboard the train would still be living on-board for several decades beyond their disappearance, thus ensuring that whatever food might be scrounged up would be inedible. Seeds that could be found and planted would still require a season or two in order to make living off the train possible. However, it is unlikely that the passengers—two generations removed from the pre-apocalyptic world—would have the skills to identify, collect, and manipulate seeds long buried beneath a frozen landscape.

Additionally, because of the cramped conditions aboard the train even the animals used for breeding food to feed the passengers would not be available in sufficient quantities to repopulate fast enough to take advantage of life outside the train. The only plants available would be those grown and nurtured onboard the train as well. There is no evidence that the elites who were planning to use Snowpiercer as a haven during the cataclysmic apocalyptic event they brought about were forward thinking enough to secure a seed bank in the event they were able to disembark at some future time. Obviously, the survivors on the train could continue to sustain themselves eating the industrial vat-grown meat that composes most of their diet, but the outside world would be inhospitable and uninhabitable likely for centuries beyond what the survivors have prepared for ensuring they can only live on or intimately connected to the train.

It is a safe bet that most of the nuclear detonations would occur on or over populated areas effectively eradicating most of humanity. Of course, there is the problem of Snowpiercer's tracks because they are laid pre-nuclear winter so they would, obviously, be traveling through the very locales that would be prime nuclear targets. This means that there is a necessary assumption that either the tracks are miraculously undamaged by multiple nuclear blasts or the tracks were intentionally laid so that they would be spared—a level of intent that betokens a complicity on the part of the elites to bring about the end of the world which is no longer entirely outside the realm of possibility. If the former, what a stroke of unimaginable luck, but if the latter, then it fits nicely with Lob's class critique. However, moving outward from each detonation site there would be massive infrastructure damage making life very difficult for anyone not on the train even if it were possible to survive long term in such conditions. The wintry conditions would set in rather quickly, but it would be possible that some small pockets of people could find ways to survive long term in bomb shelters and such given hydroponics and, perhaps, ready access to small breeding animals, like rabbits or rats—to say nothing of the possible availability of petri dish produced meat like the kind utilized on the train. And some animals suited to arctic conditions located far enough away from the blasts might persist, though diminished in number and size, with radiation induced abnormalities, but otherwise hale.

For the most part, however, Snowpiercer would be traversing a world where nuclear facilities were slowly leaking extra radiation into the irradiated world, where toxic chemicals stored close to detonation sites would be leeching into the environment, where all exposed vegetation would die and animal life unable to sustain itself would undergo a mass extinction that would dwarf the extinction of dinosaurs 65 million years ago. Which is to say, even if people figured out how to live in an irradiated world, it would not prove beneficial to get off the train for some time *after* the snow melted as the environment would be incredibly hostile to life. Of course, these conditions render life on Snowpiercer tenuous and the fragility of the environment aboard the train could be undermined by any disruption to the carefully crafted balance necessary to continue post-apocalyptic life. Lob's depiction of the end of life on Snowpiercer results not from radiation sickness, but from a disease—attributed to Proloff who is accused of bringing death with him from the abysmal conditions of the tail. The horrific nature of life and death in the pseudo-nuclear post-apocalyptic world of *The Escape* reinforces the Cold War dystopian vision Lob creates of a world at the mercy of nuclear war.

Always Begin with an Icebreaker

Snowpiercer, however, was only one train of its kind, and by the time Legrand picks up the narrative the nuclear concerns of Proloff have faded somewhat. Legrand's contribution to *Snowpiercer* features a train, like Snowpiercer—a high-tech luxury train, built for extreme weather conditions, "an icebreaker"—another train the existence of which further implicates a social elite in preplanning the destruction of the world—whose inhabitants live a life onboard their train similar to that of the unfortunate citizens of Snowpiercer. Except on Icebreaker there is more going on both in terms of how the train is managed and the interactions the residents have with the outside world.

The story is told from the perspective of Puig Vallès who, as a child, experiences the first of many "braking tests" while watching specially trained explorers leave and return from a mission *outside the train*. Puig's childhood experience takes place fifteen years prior to the present tense of the story. In Lob's story Adeline recounts her experience of boarding the train at the start of the cataclysm as a small child. She is easily in her early 20s during her adventure with Proloff, whereas Puig was likely not yet born when the apocalypse began or only an infant. If there is a little overlap in the chronology of the two it seems likely that the survivors of the original apocalyptic event have been on these trains for at least fifteen years and potentially a little longer.

Not only does the Icebreaker have a social class of explorers who are able to leave the train, at least for short intervals, ostensibly to do maintenance on the exterior of the train, they also venture farther afield scouring cities and urban centers for salvageable treasures. Their ability to leave the train when temperatures are hovering around minus 121 degrees Fahrenheit speaks volumes about how technologically advanced their suits must be which belies a high level of expectation with regard to what "extreme weather conditions" the train would be travelling through.[28] The temperature, even at minus 121 degrees, represents a significant improvement according to the reports delivered to the trains governing council, along with stability in agriculture, meat production, fertility control, gambling and entertainment throughout the train. In fact, everything appears to be going fine except that there is an increasing demand for anti-depressants and a growing fear that there will be a head-on collision with Snowpiercer.[29]

The fear of a collision with Snowpiercer is interesting because the existence of another train, built for the same purpose, but entirely unknown to the elites on Snowpiercer creates a wrinkle in the overall narrative. The Icebreaker is a substantially better and bigger train than Snowpiercer. Unlike Snowpiercer, where the accommodations were cramped, supplies limited, and the ability to produce much of anything was severely restricted, Icebreaker boasts a lot of amenities that dwarf the Snowpiercer in concept, design, and post-apocalyptic comforts. Icebreaker boasts a stocked aquarium in addition to its other agricultural luxuries, it has a shuttle system that allows for quick travel between distant cars, it has a small airplane that is used to scout the tracks for obstacles that might derail the train, virtual reality entertainment that mimics the Holodeck on *Star Trek*, and though there is still a rigid class structure, including a tail section overcrowded with the poor, there does seem to be more opportunities for social interaction and mobility than Snowpiercer could provide even if its ruling class had wanted to make such opportunities available. At one point two of the leaders aboard Icebreaker, enjoying a bit of a respite from governing in an observation deck atop the train, joke that the designers of the Icebreaker thought of everything just like those of the Snowpiercer to which one of them hollowly chuckles, "Ha ... don't make me laugh."[30] But the big secret aboard Icebreaker is that it is not going to crash into Snowpiercer because its remains—especially the engine—were already collected by the explorers during the first break test.[31]

Break Testing, Mic Check

For Legrand, *The Explorers* is the necessary link that connects his story to Lob's, but it is *The Crossing*, Legrand's second addition to the *Snowpiercer* saga, that moves the narrative beyond the final gasp of humanity Lob had envisioned. The storyline is picked up not long after Puig is shown the old Snowpiercer engine and told the tale of its confiscation by the explorers of his childhood memories. *The Crossing*, however, is not told from Puig's perspective like *The Explorers*, it is instead told from the perspective of Val Vallès née Kennel. Val, daughter of one of the council members aboard Icebreaker and now married to Puig, is one of the artistic designers developing the virtual reality stories used for entertainment.

Her story begins by sneaking into the engine room to find out what her father and husband have been hiding from her. She knows there is something important being kept from her, and Puig knows it as well, acknowledging that she deserves to know the truth, but "not to have it thrust upon her [like him]—[because] no one did."[32] The secret of Snowpiercer's existence is treated as top secret, need-to-know information, bearing dire consequences if the truth got out. This is ostensibly to use the fear of a head-on collision to justify the break tests that allow the elites to pilfer the frozen world. It hardly seems worth the trouble of hiding the reality of the Snowpiercer from the inhabitants of the Icebreaker since they are all aware that Snowpiercer exists and has gone silent. Nevertheless, the elites fear Puig, his connection to the lower classes, and what he knows, and, as a result, they want to silence him.

It is in the attempt to silence Puig that the storyline shifts from one of merely surviving aboard the train to one of attempting to establish a line of human descendants that can live outside the train. When Val and Puig first meet, toward the end of *The Explorers* story arc, she is explaining some of the aspects of life in the front of the train when they come across the nursery. Puig inquires if the babies in the nursery pods are real and Val explains that, yes, they are, as everyone in the front of the train is allowed one child. She further explains that from the ages of ten to fourteen the children are taken from their parents in an attempt "to train them to resist the cold," in the hope that within a few generations their descendants will be able to go outside.[33]

Though there is certainly some unethical science going on—Puig mentions that they use men from the tail for the experiments no one wishes to carry out on a child—and definitely some questionable parenting, the experiments do seem to be yielding results. The results are a generation of explorers that are young children, obviously forced to grow up before their time, but capable of resisting the cold better than their adult counterparts. Because of this, and their standing as the special project of the elites, they are sent outside the train to confront Puig and kill him as he investigates an explosion on the train.

Because of the conditions outside the train the survivors understand that the only hope for humankind is to develop a post-human species that is suited to the cold. That Val expresses this hope in terms of generations and the children are still wearing explorer suits it seems likely that any successes the program has had so far have been minimal if nevertheless tangible. The situation that led Puig to resume his explorer duties and venture outside the train requires the severing of the train lest the whole vehicle derail and everyone die. Once the train is successfully separated at the damaged car the biggest secret of the Icebreaker is revealed to the inhabitants of the train because of the ominous consequences now facing the survivors.

Lacking enough food to feed everyone and realizing there is limited time to decide what to do the leadership inform Puig that they have been picking up a faint signal over the airwaves in the form of music. The music, however, is coming from somewhere on the other side of the ocean. A decision is made to pursue the music in hopes of finding other survivors, but the decision comes with a no-turning-back clause because once the train leaves its tracks—which it can do with tire chains—they will not be able to turn around nor, presumably, find the tracks again. Up to this point in *The Crossing* all the depictions of the world outside the train have been typical of the snowbound, frozen world that has consumed the entire planet. With the exception of a minimally detailed explorer visit to a city there are no meaningful engagements with life outside the train that were not shrouded in perpetual darkness and ice similar to the world Snowpiercer traversed. Once Icebreaker reaches its jumping off point for its transoceanic journey, a harbor and shipyard on some unnamed coast, there is once again a depiction of the world that bears little resemblance to the world that has been described by the ethereal narrator that continually reinforces the bleak and unforgiving nature of the wintry apocalypse. The harbor, much like the rail yard Snowpiercer passes through, is remarkably discernable despite having been abandoned for decades to a perpetual winter.

Perpetual Winter, Eternal Summer

The crossing is a slow and arduous journey made worse by the fact that the train is continually slowing down as a result of being off of its tracks. The journey takes days and as the time at sea passes the world outside the train becomes a flat surface, and endless sea of white in all directions, except for one glimpse of a ship that is frozen in the process of sinking, rudder in the air, its other half silently submerged beneath the ice.[34] It is a curious sight to be sure and one that is so inexplicably out of place that it is a disruption to the story itself. How does a boat, afloat on the ocean at the time the apocalypse began, sink so slowly that an apocalyptic winter could freeze the ocean and capture it before it could find its way all the down to the bottom of Davy Jones's Locker? It would take an awesomely powerful, incredibly cold, and terrifyingly fast storm to be able to pull off something like that—one that surpasses what a nuclear winter could produce even if the ocean were bombed.

The average freezing temperature of saltwater is about four degrees lower than freshwater, but the problem faced by a scenario like this is that as the ocean froze it would displace more salt into the water below the forming surface ice. As the salinity of the water increases its freezing point continues to lower. So, even if the oceans only froze, say, 10 meters deep, the question remains whether that would be enough ice to hold an ocean liner suspended mid-sink and, if so, how long would that take versus how powerful would the storm have to be to achieve such a feat? It is another obvious misrepresentation of the post-apocalyptic world and though it is a misrepresentation that does not necessarily make the apocalypse seem more livable, the more inaccurate the world, the more likely people are to mis-imagine the realities of life in the aftermath of nuclear annihilation.

While it goes without saying that the problem of mis-imagining post-apocalyptic life could easily be applied to every scenario central to my analysis in this text, the underlying issue is a bit more nuanced. The problem is one of closure which Scott McCloud defines as the phenomenon of observing the parts but perceiving the whole.[35] Where the parts of a graphically depicted story, no matter how minute, are misrepresented, then it becomes increasingly difficult to perceive an actual "big picture" and this problem is exacerbated when the depictions available to us as readers are not merely inaccurate, but, in some cases, positive. Inevitably, we are going to draw from our past experiences to achieve closure even though we all understand that we have not experienced an apocalyptic aftermath.

Forty days into the crossing food is running short, potable water is nearly gone, and the terrain of the ocean is becoming rougher as the Icebreaker must employ its front mounted guns to blast through obstacles—water frozen mid-wave—that the slowing train can no longer plow through. The instability and increasing hopelessness of life on the train coupled with the dimming hopes of finding other survivors provide the buildup for the conclusion of Legrand's story. As the few remaining humans face the reality that they may not survive their ocean voyage Puig takes the airplane out one more time to scout the area. His brief flight brings him face to face with an aircraft carrier whose aimless drifting was brought to a halt long ago by the iced ocean. Though the light they see on the carrier is attributed to the "last gasps of an atomic battery" it still has the juice to move the automatic defense system which fires two missiles at the train reducing it to one-tenth its original size.[36]

Eventually, the train makes its way to an icy mountain and when it comes to a stop, unable to move further, Puig ventures out and up the mountain toward the source of the music. In a final nod to the unpredictability of the story's use of the environment the music is emanating from a broadcast station that is mostly buried by snow, but not entirely. In fact, the mountain of snow that Puig climbs up has not managed to fully engulf the door to the facility in spite of being subjected to more than a decade of snow that has not been deterred in its accumulation. Compare this to the aircraft carrier frozen in place on the open ocean. A Nimitz-class aircraft carrier is taller than a 20-story building so, allowing that a single story is 12 feet, the rough height of a Nimitz-class carrier would be approximately 240 feet. Even assuming a smaller carrier at 200 feet tall that would still be the equivalent of a 17-story building.

Even using the smaller vessels height and allowing that the draft of the fully loaded ship puts one-third of it below the waterline there would still be about 135 feet of ship above the water. That is 135 feet of steel broadside that has supposedly been catching snow and ice for more than a decade, but when the ship is shown up close it looks like it was just recently frozen in place.[37] Of course, it is entirely possible that the aircraft carrier only recently came to a halt as it would take a long time for the oceans to freeze solidly enough to immobilize a ship of that size. Moreover, the aircraft carrier is a considerable distance past the other ship that Icebreaker passes. The obvious problem is that the same ocean that froze fast enough to capture one ship mid-sinking failed to halt the aircraft carrier; but if both ships stopped moving at approximately the same time, even granting severe wind, the aircraft carrier should have been, especially its decks, covered in substantially more snow—perhaps even enough to thwart the automated defense system.

Granting that the size of the aircraft carrier allowed it to move about for a considerable amount of time after smaller vessels were frozen in place it should still be encased in quite a bit of subsequent ice and snow. However, the aircraft carrier barely has any snow or ice burying it—and none clogging up the deck mounted missile launchers. Still, when Puig arrives at his musical destination the snow has turned the landscape into a series of mountains ostensibly hiding all the buildings in the area except what must be a very tall building that was operating, at least partially, as an automated broadcast station hooked to a power source that could continue to broadcast for fifteen or more years without human maintenance—perhaps another atomic battery.

Upon finding all the people in the broadcast booth long since frozen Puig despairs that he has led them all to their deaths. And considering that they are now living in a train stranded in the middle of nowhere, with resources that are rapidly dwindling to nothing, in a world where contact with the outside air entails immediate death, it is quite likely that he has. Legrand finishes his continuation of *Snowpiercer* on the same sour note that Lob chose to end his. The inhabitants of Icebreaker never reveal what caused the apocalypse, nor do they speculate about the apocalypse in general, they are instead focused on surviving, improving their lives, and overcoming the apocalypse by engineering cold resistant children. While this orientation to the world provides a more optimistic and hopeful worldview, especially given how well-appointed Icebreaker is compared to Snowpiercer, it ends with the same hopelessness that Proloff expressed as he lived alone as the last person on earth in Lob's particularly harrowing conclusion. But the lack of reflection on the apocalypse leaves it an open question whether the leadership aboard Icebreaker knew things about the apocalypse that the inhabitants of Snowpiercer did not know. Having made no such concessions, it seems only fair to assume that the survivors in Legrand's story arc share the same beliefs about the apocalypse as the survivors in Lob's original tale. If that is the case it certainly makes Bocquet's contribution all the more interesting.

A Tune That Never Stops

Although Bocquet's story arc literally picks up precisely where Legrand's ended Bocquet begins with a concise recounting of the *Snowpiercer* saga that begins to reshape the narrative even more noticeably. The nuclear war presumption gives way to a more mysterious glaciation and the train becomes Snowpiercer again as if the Icebreaker, once mentioned as Snowpiercer 2, is the train that has been traveling the tracks this whole time. Also, the explorers are now presented as trained and equipped to survive the cold though Puig made it unquestionably clear in Legrand's construction of *Snowpiercer* that the men chosen to be as explorers were expendable.

Bocquet's description makes them seem indispensable and valuable. He describes the events that have led the survivors to the brink of extinction.

> A climactic cataclysm has plunged our planet into a sudden glaciation. The luckiest people died immediately. The few remaining ones embarked on a train that never stops: the Snowpiercer. After decades of wandering, violence and political struggles, the train picks up a radio signal from across the ocean. Mounted on makeshift caterpillars, it crosses the frozen area, sacrificing on the way most of its cars and passengers. At the end of the journey, four explorers—people trained and equipped to survive the cold—climb a mountain to the very source of the music, in the hope of finding survivors. They only find frozen corpses. And a transmitter, playing a tune that never stops...[38]

The first scene in *Terminus* is a panoramic view of the frozen apocalyptic landscape, dominated by the mysterious mountain, the Icebreaker train, and the lonely notes of the automated sirens call that led the survivors to their doom.[39] The icy hopelessness Legrand concluded with quickly turns into a survivalism as determined and rigid as the world is frigid in Bocquet's narrative. Defeated, Puig and his compatriots decide to return to the train. Val, however, exhorts him to stay outside the train and continue searching for anything that might justify their having traveled across the ocean, especially the energy source that made transmitting the music possible.

She says, "all that's left of humanity is crammed into these ten carriages. People are injured, ill. We're overpopulated, we have no food, no sick bay, the stokehold is suffocating … there's nothing in the train that can save us."[40] Val accurately captures the nature of their situation but quickly decides that, before succumbing to the reality they are facing, every option for survival must be exhausted. So, she instructs Puig to go and find the energy source making the transmitter operational because they "didn't cross the ocean to find salvation" and the electricity juicing the transmitter is their "only hope."[41]

Sure enough, with a little digging the explorers find a door that opens to reveal artificial light. Soon, however, they realize that they are not at the bottom of something, that is, in a basement, but rather, they are at the top of a skyscraper. A couple of things immediately clash with the way Legrand presented the end of the oceanic journey. First, why would the explorers think they were heading into the first floor or basement of a building when they just climbed a mountain of snow to get to the transmitter room? Beyond that, however, is the revelation that they are not only not entering a lower level they are on the forty-second floor of a skyscraper.[42] How is it that a skyscraper, at least 42 stories high, easily more than 500 feet tall, is completely encased in snow and ice but the aircraft carrier, a structure with a maximal height of 240 feet, had minimal snowy buildup? Even making allowances for the fact that the building is located in a cityscape where snow would likely accumulate faster than it would against the only obstacle in an open area occupied by the aircraft carrier, all the other boats previously depicted docked in harbors or adrift at sea should be covered in a lot more snow. If it were not already obvious that the wintry apocalypse was not affecting the world in the ways that a decades long blizzard should be, then at this moment it most certainly is apparent.

While Puig and the explorers apprehensively investigate the electrically operational building the remaining leadership aboard Icebreaker debates what to do next. While they are discussing these matters, including what to do about the peasant uprising that is taking place, three people peer into the trains engine control room as though they are attempting to spy on the leadership. Astonishment mixed with fear grips the counselors as they exclaim in unison, "how can they be alive," and "no one can survive outside without an ice suit."[43] And, though Val wants to let them in regardless of who they are, counselor Brady shoots out the window of the train attempting to kill them. Breaking the glass exposes them to the cold, even behind the security shutter, and the temperature begins to fall rapidly. Fortunately, instead of killing the interlopers, they make their way into the engine—how they do so is never explained—and it is revealed that they are children citizen-passengers of Icebreaker.

In their explanation of who they are the children provide more information on the adaptation program that Val briefly explained to Puig when they first met. One of the children explains that they are "part of the adaptation program [designed] to adapt mankind to the climate." He further explains that they are "trained to resist the cold," which is supposed to alter their DNA to make them cold resistant.[44] Though the children are not hybrids like Gus and his friends in *Sweet Tooth* the use of mutated or genetically altered children to overcome the apocalypse is clearly at the heart of both Lemire's and Bocquet's stories and offers a glimpse into how the human element may only emerge from the apocalyptic after undergoing some sort of alteration to the species as the only viable option for successfully surviving a cataclysm. What is possible in the frozen laboratories of Alaska and Snowpiercer, however, is not the end of the possibilities in Bocquet's *Snowpiercer*.

Rat People: A Mutation Story

Using an elevator shaft, Puig and the explorers descend to the bottom of the building and discover a city with the lights still on and a train station separated from the outside by a wall of ice that Icebreaker can easily break through to enter the underground haven. *Terminus*, at this point, takes on a narrative structure that is radically different than its predecessors' vision of the *Snowpiercer* world. After crashing through the ice and entering the underground station the temperature is warm enough to allow the passengers to disembark without ice suits and, for the first time in decades, the world of *Snowpiercer* is bigger than the confines of a train. The celebration of finding a habitable space quickly gives way to a continuation of the political uprising which now has the momentum of not requiring a hierarchy imposed by the design of the train and the long suffering second and third-class citizen-passengers are committed to ushering in a new era of democracy. Toward that end one counselor is brutally murdered and crucified on the cowcatcher of the train while Puig is arrested and put in prison awaiting trial.

Meanwhile, in the process of exploring their new home, the children of the adaption program cross paths with a golden retriever and decide to follow it.[45] Discovered to be missing, Val is sent to locate and return with the children if she wants to act as Puig's lawyer at his upcoming trial. Over the next pages there is nothing to indicate what is going on inside the station as the action focuses on a forgotten Puig who is painfully languishing in prison—which is nothing more than a coffin-sized lockbox. By the time an emaciated and nearly dead Puig is discovered by the train's engineer he is told that everyone has been gone for twenty-nine days.[46] As it turns out, instead of finding the children, Val and her search party find the inhabitants of the station as they wander its various corridors and levels.

That the apocalypse has been on-going for only about two decades means that everyone on the train that was older than, say, five, when they boarded the train ought to remember what it was like to live in the pre-apocalyptic world. Yet, as Val and her team search the station they are bewildered by the experience and while it is likely that some of the people on the train had never experienced automatic lights, or escalators, it is unlikely, and even without having experience of those things it is likely they would know what they were.[47] More to the point, when the group comes across an ad for toothpaste two older men, men who were definitely alive pre-apocalypse, do not know what it is and, worse, surmise it is some type of pre-striped paint causing one of the men to wonderingly muse, "Shit. Civilization, man."[48] The idea that the survivors, especially older survivors, have no recollection of pre-apocalyptic society is absurd, but to have the survivors behave with such puzzlement is a step too far. Bocquet's narrative represents a radical break with the chronology of both Lob's and Legrand's stories, but a disruption in the timeline is not the only unsubtle break Bocquet introduces into the story.

Once Val and her rescue team make contact with the inhabitants of the station they, along with all the disembarked citizen-passengers, are led to an operational movie theater that has been repurposed as a holding facility. While there they are provided with fresh fruits, hot showers, and clean clothes. Obviously skeptical about their incredibly good fortune the survivors are slow to come around to the station inhabitants' ways of doing things. Eventually, Laura Lewis, the leader of the democratic rebellion, appears on the theater stage to announce that they have finally reached their destination. More than that, she informs her fellow travelers that there were *ten* Snowpiercers and they are the seventh to end their journey at that station; and, since the other three are missing, they are likely the last one.[49]

One of those "missing" Snowpiercers is obviously the train from Lob's original story, which means there are two other trains somewhere traveling the planet, with potentially hundreds, if not thousands, of survivors. Icebreaker, prior to being damaged by the missiles from the aircraft carrier, had approximately 1,500 people onboard.[50] If half that number was lost to the unfortunate run-in with the aircraft carrier, there would still be 750 people on Icebreaker. Prior to jumping its rails Icebreaker was the same size as Snowpiercer, 1,001 carriages long, which means that each train, each of the ten known Snowpiercers, could carry several thousand passengers comfortably. If the survivors were crammed in, then it is likely that the number was higher for some trains than others, but contrary to Lob's assertion that the last remnants of humanity were aboard the Snowpiercer, there are actually potentially thousands of survivors in the frozen wastes of the post-apocalyptic world of *Snowpiercer*. Again, Bocquet conjures up a radical break with the original narrative which, like the other post-apocalyptic dystopias discussed thus far, introduces a hopefulness that undermines the reality of a nuclear winter style apocalypse.

Up to this point the long-buried city Puig and his companions originally stumbled upon has been depicted as a skyscraper connected to a shopping mall above a subway or light rail station. That a group of people could keep a regular old building operational for two decades, during an apocalypse, and not just manage to get by, but to thrive, definitely strikes a nonsensical chord that does not resonate with the rest of the *Snowpiercer* narrative—until the passenger-survivors are brought out of quarantine. What the passengers have located is Future Land, a nuclear research facility designed as an amusement park with the lofty pre-apocalypse goal of making interstellar travel a real possibility. Future Land is an underground "autonomous city" which used admission sales to fund research initiatives like self-sufficient farming, cutting edge technology for provisioning water, breathable air, electricity, heat and light, as well as anti-aging technologies, all of which were being designed to tackle the biggest problems facing humans with the ambition of traveling to far off planets in deep, dark space.[51]

Obviously, with space age technologies and a high population of scientists among the residents of Future Land, it is no surprise that the inhabitants were able to achieve what they did in spite of the apocalyptic conditions engulfing the world above ground. Future Land is an atomic age Noah's Ark with a zoo that houses the last remaining specimens of many animals, bamboo forests that provide the material necessary to make a multiplicity of products for personal and industrial uses, a large scale farming operation, an economy that allows for the trade of goods, ample employment that draws upon the special skills of each member, a sewage system, exercise equipment, a museum, a working rollercoaster, and sufficient housing to allow everyone to enjoy a little personal property even if they do not all have access to private space. The giant LED high definition walls even project a beautiful, if repetitive, sunset each night which makes for a singularly pleasing aesthetic moment in an underground existence. Of course, as with all things too good to be true, Future Land has a serious problem.

The Promised Land Is Lost

Puig, having escaped capture and quarantine by virtue of having been incarcerated in the train, has been on the loose, unbeknownst to the inhabitants of Future Land, and carefully surveying the compound, deciphering its many mysteries. When he finally intercepts Tom, the Icebreaker's radarist and a former counselor, as he heads for his new job assignment Puig explains that Future Land is a nuclear facility. Alarmingly, Puig informs Tom that the nuclear facility is faulty and explains to the radarist that his new assignment will be his certain death because he will be repairing the failing nuclear facility that sits above the amusement park-cum-research facility utopia. Clandestinely gathering his former explorer compatriots Puig infiltrates the heart of the Future Land compound—partly to expose the truth behind the community, but mostly to locate and rescue Val who, being pregnant, has been inexplicably imprisoned deep in the underbelly of Future Land.

During their infiltration of the research facility the realities of living in concert with nuclear energy are made clear as malformed children, mad scientists, and devoted cultists have been hiding behind a façade much like the Great Wizard of Oz. The scientists, like their counterparts on Icebreaker, are working to develop a human child that can live and thrive in the apocalyptic world. Obviously, this is a different approach to post-human mutations than what was taking place in *Sweet Tooth*, not the least because rather than performing the experiments out of mere scientific curiosity, the scientists in *Terminus* are committed to the belief that they are doing what they are doing "for the greater good."[52] That is, having turned their cosmological dreams into efforts to prevent the extinction of humankind they now think of themselves as the saviors of humanity, or did, until an earthquake destroyed the foundation of the nuclear facility leading to radiation induced illnesses and the near destruction of life at Future Land.[53]

It is no surprise that the desire to create the "new man" should follow from the overall story arc of *Snowpiercer*. In a world destroyed by nuclear war, or less hyperbolically, a world destroyed by human actions, that results in the conditions commonly associated with nuclear winter, the possibility that a small group of survivors, even a community as potentially large as the survivors aboard the various trains and Future Land, would not be able to outlive the environmental effects on the world. The conditions that exist outside of the trains and Future Land are too harsh and, per the ethereal narrator of the story, seemingly eternal. It would be impossible for humans to survive long-term, not to mention longer than the frozen hellscape that has been created, without some modification to what it means to be a human. The inherent problem in the logic of Future Lands' leaders is that they are committed to the idea that the source of their problems, nuclear age science, is also the key to their salvation, but it is not.

Bocquet breaks with the bleak narrative of Lob and Legrand in a couple of ways that establishes *Terminus* as an entirely different slant on the apocalypse—one that has to end on a positive note. First, unlike his predecessors,

Bocquet, along with Rochette, gradually changes the story's presentation from the stark black and white of Lob and Legrand to one that is increasingly color filled. Second, the idea that people have been living off the trains provides an additional glimmer of hope to the narrative. Consider how differently it would be for the inhabitants of the train to find out there were ten Snowpiercers, of which seven were accounted for, and three were missing, in Lob's or Legrand's narratives. The hopelessness that gripped Lob's inhabitants would undoubtedly be compounded by the idea of the lost coupled with the realization that there were fewer and fewer people. However, in Bocquet's narrative it becomes a source of hope that they are out there, perhaps having already been rescued by some other group living underground, or having established a haven of their own somewhere where the weather was not apocalyptically cold. Future Land's existence becomes a blueprint for believing that life in the apocalypse is possible and, further, that life after the apocalypse is also possible.

Finally, once Puig rescues Val, and the people living in Future Land are given the choice to go with them or stay behind, a choice fully informed by what the scientists are doing, what is happening to the nuclear facility, and what life will be like on the repaired Icebreaker, Snowpiercer 2 once again heads out into the frozen waste. Leaving turns out to be a wise decision for Puig and the people who chose to rejoin the sojourn of Snowpiercer. After a bloody departure from Future Land, the survivors are shown orgiastically feasting on the body of a whale pulled up from beneath the ocean ice. The story has progressed some fifty years since Puig and Val led the exodus from Future Land, and though their time aboard the refitted Icebreaker goes unmentioned, the survivors now live in the train as if it were a cave. They draw upon its walls to mark their history, and a long history they have lived since leaving Future Land, as one of the adults tells a child, "no one really knows how long it's been. [We have] lost count. [But] some ancestors think that the great cold started a hundred years ago."[54]

Here, again, there is a narrative problem. The time frame of *Snowpiercer*, from the beginning of the apocalypse to the moment Puig and Val leave Future Land, is, generously, two decades, maybe twenty-five years *at the most* and it is more likely that the amount of time that has passed is closer to fifteen or eighteen years. Nevertheless, Bocquet's anonymous adult tells the excited children it has been maybe a hundred years since the cold started and refers to "ancestors" but Puig and Val are still alive. Granted, they are incredibly old by this time, each appearing to be well into old age, but they would have to be nearing 90 for the chronology to align. And, to be fair, that is possible, but then there would be no need to speak of them as ancestors nor to have "lost count" of the years since the apocalypse began. Puig is blind at this point in the story and Val describes the world to him as they are taken by an excited child to witness the "giant fish" the fishermen have pulled from the ocean.⁵⁵

There is no explanation for how the train, which required tracks to generate the energy to keep the train moving, and which was continually slowing down as they originally crossed the ocean in search of the mysterious music, was able to leave the station and journey anywhere substantially far away from Future Land. Of course, it is entirely possible that the train has come to a standstill on the ocean, explaining why there are fishermen and the train is treated like a cave dwelling. But if that is the case, then the survivors on Icebreaker's post–Future Land journey are luckier than they perhaps know because they are catching killer whales within walking distance of dryland. After taking in the spectacle of the killer whale's death and consumption Puig asks Val to take him to a higher elevation that he may see the grass growing again. Val escorts Puig to the top of a nearby hill and the subsequent depictions of the world are black and white, desolate and bleak, as forbidding as any of the depictions of the landscape in any of the versions of the world throughout *Snowpiercer*.

Puig, unable to see for himself what the world looks like and reconciled to the fact that his death is near, asks Val to describe what she sees from their vantage point. As he lays in her arms she describes the world as green and covered in "grass and leaves and trees."⁵⁶ Just as Bocquet and Rochette have done throughout the *Terminus* storyline, the black and white of the world slowly gives way to more and more color as Val details the white expanse of the apocalyptic landscape with imaginary colors. The last words spoken in *Snowpiercer* belong to Val as she tells Puig, in his last moments, that there are "flowers everywhere ... red and yellow and purple and pink. And blue ones."⁵⁷ On their own, Val's words, even allowing for the bluing of the sky as the story ends, seem to retain a bit of the utter hopelessness of the post-apocalypse that Lob captured in Proloff's last days because she is lying to Puig. But, right at the end of the story, in a corner of the world out of Val's sight, away from the survivors feasting on the whale, blue flowers are seen growing in a patch of dirt exposed by melting snow.⁵⁸

The Absolute Effacement of All Possible Traces, but Perhaps Not

Snowpiercer represents the worst-case scenario for a human caused apocalypse and, again, not only does humankind manage to survive, but so does other flora and fauna—for, miraculously, upwards of a hundred years in spite of an icy apocalypse. However, the plight of the survivors is not likely to improve just because snow is melting and flowers are blooming. Once the passengers of Snowpiercer were able to fully disembark and live in the world away from the train they would find themselves at the beginning of a nuclear summer. In the event of a nuclear winter, it would inevitably be followed by a nuclear summer caused by the decay of all the organic matter that had hitherto been frozen. One can readily see that on or off the train the social critique presented by *Snowpiercer* is of fundamental importance; but, if the critique is going to make sense, if it is going to be maximally effective as a persuasive argument for social and environmental responsibility, then the critique must be situated in the context of what exactly a nuclear winter entails.

The limited resources of the train would seem like an abundance by comparison to what is waiting under the snow and ice and as such the conditions of the train seem even more dreadful. In reality the inhabitants of the train would likely have to continue living according to the rationing principles of Snowpiercer. One killer whale does not a cornucopia make and though the story ends on a hopeful note there is no reason to be hopeful about the prospects of the people still struggling for survival in the *Snowpiercer* apocalypse.

Where *Snowpiercer* presents cogent arguments for the necessity of resource distribution, egalitarian living conditions, and respect for all persons—or at least giving each person equal weight along utilitarian lines before, say, cutting hundreds of cars loose to lighten the load—and the need for social cooperation rather than the tyranny of a privileged elite, the story is told as if life must continue on Snowpiercer *forever*, and well it might because nuclear winters inevitably turn into nuclear summers and there is no way of knowing how the ebb and flow of seasonal change might occur or affect the survivors after nuclear winter gives way to its summery counterpart. Yet, as a reader of the story, compelled to believe that eventually the wintry conditions will subside why should one assume the snow will melt away to reveal a new Eden. Why should that be the assumption? Even if it were the case that the snowy apocalypse ended short of one hundred years after the bomb, even if radiation and freezing temperatures did not kill every living thing not on the numerous trains or housed in underground safe havens, the world post post-apocalypse would be as unforgiving and hostile to the survivors as the frozen landscape they traveled through and subsisted in for a century. They would not find a new Eden under the snow, but rather, a new form of hell they would have to figure out how to survive.

That life would continue on, regardless of the hell on earth conditions that were wrought upon it, is, in reality, undeniable barring a disease or cosmic event that kills everything. Though how life would continue is questionable, the larger concern is that life would have to continue in conditions that are apocalyptic, but not as immediately devastating as the previous examples make it seem. It does fly in the face of the apocalyptic as a concept to suggest that it would happen slowly but given the reality of global climate shift as the most likely way that the apocalypse will happen examining life in a climate shift setting seems the most reasonable way to determine what life in a post-apocalyptic future will be like. To that end, our focus now shifts to Brian Wood's conception of environmental upheaval in *The Massive*.

6

The Massive and Life on a Warming Planet

Welcome to the Tropic of Chaos

It ought to be clear at this point that whatever world is left in the post-apocalyptic environment will be incredibly hostile to survivors and the deadliness of the environment will be a long-lasting reality that would encumber any efforts to re-establish civilized society. Attempts to usher in an era of stability and peace would be thwarted continuously not only by an environment that would be heaping suffering upon suffering onto survivors, but by the shenanigans of humans attempting to exploit the lawlessness that would exist as a by-product of any post-apocalyptic reality. This is not to suggest that life would not continue, nor is it realistic to believe that certain human elements of the world would not continue on or eventually re-emerge in the aftermath of the apocalypse. As Jamais Cascio flatly asserts, "the 'end of the world' isn't. It's just the beginning of what's next."[1] Cascio's position only makes sense when it is held relative to a future like the one presented in *The Massive*.[2]

As such, the question that, heretofore, remains unanswered is what life would be like in a dysto-apocalyptic future that did not begin with a bang, so to speak, where all life was not obliterated in an instant. Rather, the most persistent, complex, and likely, threat to a human future is to be found in a future birthed by global climate shift. Wood's telling of the slow, but persistent, breakdown of the social order using global climate shift as the engine develops the apocalypse as a process, a series of events, that involves a combination of a potentially destructive agent from the natural and technological spheres along with a population in socially produced conditions of vulnerability. The development of the process and subsequent events produce damage or loss to physical facilities and to major social-organizational components of various communities, to the extent that the essential functions of the group are interrupted or destroyed. When this happens individual and group distress and social disorganization of varying severity follow.[3]

The Massive, unlike its previously analyzed counterparts, imagines the world in the ongoing throes of the most likely, human-made, global catastrophe, and presents a near-future suffering the ill-effects of global warming—or, at the least, anthropogenic causes. A combination of climate change and other pre-apocalypse conditions linked to industrialized, capitalistically oriented practices, seem to be the culprit throughout much of the story.[4] It is important to connect these two things because climate change, when it arrives in its apocalyptic regalia, will arrive in a world primed for crisis. The current and impending dislocations of climate change, argues Christian Parenti, intersect with the already-existing crisis of poverty and violence in what he calls catastrophic convergence. Catastrophic convergence is not merely several disasters happening simultaneously, but rather, the problems compounding and amplifying one another and expressing themselves through each other.[5] *The Massive* does an excellent job, most of the time, of capturing exactly this relationship between pre-existing social conditions and apocalyptic scale global climate shift.

Of the many things that make *The Massive* interesting, and certainly one of the aspects that sets it apart from the other examples examined herein, is that the story is told from a non–Western perspective. The result is that, as I mentioned previously, the examples derived from data about the United States will need to be extrapolated onto a global conception. Not just to imagine, say, what the problems caused from hazardous materials might look like in another locale, which is important, but to begin to see the problems as interconnected and global in scope. Conveniently, in the early telling of Wood's story, the United States has "gone dark" and little to no information about the country is available to the characters in the story. So, drawing upon the analysis from previous chapters, the natural conclusions about the condition of the United States should be easily discernible. How the rest of the world fares is the focal point of Wood's story.

From Birth to Ruin

In the context of the world created in *The Massive* inexplicable environmental upheavals have been on-going for a year all over the world leading up to the opening of the story. The setting of the prequel, its chronology, and its narrative construction are such that the upheavals that underpin the concerns of *The Massive* take place much later than the vignettes which comprise the prequel, so it is clear the stories are substantially chronologically separate. The prequel is meant to fill the in the backgrounds of the primary characters and situate them in a pre-apocalyptic struggle of environmental protection. Focusing on their efforts and the eventual formulation of the core principles

6. The Massive *and Life on a Warming Planet*   129

of the Ninth Wave—the direct-action environmental activist group at the heart of the story.

Wood describes the story of *The Massive* as the recounting of the difficulties faced by his environmentalists as they attempt to adapt to, and define themselves after, the events of "the Crash," as the apocalypse is known. The portrayal of the group is meant to capture the two-fold nature of life post–Crash: one the one hand, their travails are meant to document and bear witness to the unfolding of the apocalypse and, on the other hand, their experiences are meant to highlight the uniquenesses and generalities that exist in a world fraught with "perseverance and survival … activism and compromise … conflict and collectivism, love, death, and *a stubborn hope for a better tomorrow*."[6]

Perhaps it is because the Crash is ongoing that the characters believe they will eventually emerge into a "tomorrow" that will be the beginning of something better. The apocalypse is, if nothing else, as bad as things can get, so in some sense the post-apocalypse must be about circumstances getting better eventually. To make the case that this so, however, there must be something left to work with and, unlike the stories in the previous chapters, there is a sense that people are working to make do until they can make better. And, unlike other stories that rely on sudden onset apocalypse, it may be the slow and steady pace of Global Climate Shift that allows them to do just that.

Prior to the Crash the Ninth Wave was committed to non-violent, direct intervention, eco-terrorism as the primary means for stopping environmental degradation and consciousness raising. These goals guide the group and each mission undertaken, whether they are impeding whalers or sabotaging hazardous chemical manufacturers, is meant to hone their message and their morality. The problems that the Ninth Wave tackle in the pre–Crash world are familiar, recognizable problems presently facing the global community. Industrialized destruction protected by government corruption, struggling against deforestation, working to save threatened and endangered species whose habitats are being destroyed by development and human greed. Throughout the prequel stories Ninth Wave grows and matures both tactically and philosophically.

Perhaps the most important realization of all their growth comes when the group must decide if the organization has obligations to people, and, if so, what their humanitarian obligations might be. Callum Israel, leader of Ninth Wave, argues, "Ninth Wave can't only be about wildlife and the ecosystem [because] humans live on this planet too."[7] Though *The Massive* is decidedly a story about environmental destruction and an apocalyptic future Wood tries to incorporate the human element into the narrative structure, but outside of Ninth Wave, not many of the story's characters seem to realize they have to learn to live in concert with nature as *a part of nature instead of apart from nature.*

From Traumatic Event to Mythology

The pre–Crash world that Ninth Wave is trying to save closely resembles the real world replete with its contradictions and difficulties and degradations. When Callum Israel's world starts to fall apart it happens the way one would expect it to, just like it is happening today. Isolated incidents which, no matter how tragic, being managed and overcome through typical disaster relief efforts at the local, state, and global levels. In fact, Wood works a lot of real-world issues into the narrative of *The Massive* giving it an eerie parity to the reality of the way things currently are globally. Disaster reporter Johnny Colt draws comparisons between the real-world disasters he has experienced to the way *The Massive* captures the feel of catastrophe. Referencing the Deepwater Horizon oil spill, the human-caused cholera epidemic in Haiti, and the tsunami that destroyed Japan's Fukushima nuclear reactor complex, all of which Colt experienced as a reporter, Colt claims, "*The Massive* is no far-fetched sci-fi thriller [but] a slate of dramatic and perhaps inevitable climate-change scenarios."[8]

By doing so, the factors in Wood's story that get worse, or develop from already bad existing conditions, take on an urgency that stories like *The Walking Dead* or *The Last Man* cannot really capture. The innocuous nature of the Crash's early occurrences overshadows the truth that we are not inoculated against catastrophe and, in the event something like the Crash were to occur, unlike its zombie counterparts, seemingly localized problems can quickly become global concerns for which we, collectively, are not prepared. The Crash is not explained at the outset of the story, but rather, details are provided piecemeal, spliced into and spread throughout the early vignettes that detail the life and work of Ninth Wave. Large frames in the graphic novel are used to depict particularly harrowing instances of the Crash, seemingly disconnected from each other or a single cause, but linked together as contributions to a larger disaster narrative by their calendar dates and Wood's explanations.

The first in a yearlong series of catastrophic events occurs on January 4 as the Cook Islands are subjected to a storm that defies any formal, or informal, description. The destruction of life is near total as whole landmasses disappear.[9] A mere month later additional severe storms rock the North Atlantic and decimate shipping around Nova Scotia and, unlike the Cook Island storms which are shown to be angry thunderclouds moving in over the island, the storms of the North Atlantic are presented as forceful enough to capsize a large container ship.[10] In May, an underwater landslide destroys several oil platforms simultaneously, likely the result of seismic activity, and, because the platforms could not be properly secured in the aftermath, the resulting fires continue to rage, permanently darkening skies up and down the California coast.[11] The same month an underwater earthquake rips its way north from the Macclesfield Bank redrawing the coastlines of several Asian cities.[12] Each of these disasters is terrifying in their own right and, should something similar happen, there would be an immediate global response to address the disaster, but should similar events occur nearly simultaneously the ability of the global community to react adequately would leave much of the devastation in place.

However, these are not isolated events and, while these major disasters are wreaking havoc on the underwater geography of the ocean, the shifts begin altering the wave patterns of the ocean while massive numbers of Bluefin tuna wash up dead on beaches along the North African coast.[13] Birds mysteriously fall dead from the sky in droves in Italy.[14] Adding mayhem to misery terrorist groups are still bombing government infrastructure, ostensibly because of worsening economic conditions. This is all exacerbated by incredible losses of marine life that disable fishing markets as well as corresponding drops in energy production as wind patterns shift as the surface temperature of the ocean rises.[15]

Rising temperatures in the ocean are fed by continued glacial melt in Antarctica which, much like today, continues at an alarming rate. Wood's vision in this respect is simultaneously remarkable for its prediction and unremarkable because it is merely the necessary logical outcome of scientific warnings about Antarctica since 1995. Scientists are now concerned that a piece of the Larson C ice shelf, as big as the state of Delaware, could break off soon. If so, the ice shelf will be at its smallest size in recorded history.[16]

Later in the year of the Crash, seismic activity around the world is becoming so frequent that landslides, earthquakes, volcanic eruptions, and tsunamis were becoming frighteningly common and increasingly destructive.[17] This includes a two-week stretch in November that produces enough tsunamis worldwide to completely undermine the global economy, destabilize numerous governments, and leave thousands of people dead.[18] Rising sea-levels have fundamentally altered shorelines, national boundaries, and inland geographies as major cities are now partially, or in some cases fully, submerged—this is the reality of Hong Kong in late December when Ninth Wave limps into its makeshift port and finds it swimming in ten-stories of water.[19] Which is to be expected when the oceans reach a level capable of swallowing the entire island of Taiwan.[20] But, even in the event that rising sea levels did not reach the point where Taiwan was submerged, or, in the context of the graphic novel explanation, if Taiwan were to have been hit by a tsunami prior to being submerged Hong Kong's harbor would be untenable.

As recently as 2010 Taiwanese officials expressed concerns about rising sea levels that were bringing more regular flooding into areas farther inland. Not only does this flooding taint the land with salt deposits, but the ebb and flow of the ocean at these higher levels destroys infrastructure, shorelines, and livelihoods.[21] Taiwan's western coastal plain is home to a string of cities, several industry zones, *three nuclear power plants and a petrochemical complex* all of which are at risk to rising seas.[22] If, while watching the seas rise, a tsunami caught the Taiwanese people off their guard, like the tsunami that crashed into Fukushima, the result would be beyond disastrous. With a landmass less than a tenth the size of their northern neighbor Japan it is unclear, if it would even be possible, for Taiwan to even minimally manage the disaster.[23]

Along with these early instances of catastrophe, each of which is presented as being a strange anomaly locatable to its own area of destruction, other incidents fill up the Crash calendar. Heavy snows that leave small towns isolated and imperiled, the drying up of the Suez Canal, which strands the last boats whose captains braved the dwindling waters attempting to make the dangerous crossing. Satellites fall from the sky as the earth's electromagnetic field shifts inexplicably. High amounts of methane burbling up from the bottom of the ocean rapidly reduces the surface tension of the water, sinking an entire American battle group.[24] An unknown cause results in the detonation of a British nuclear submarine in the Strait of Magellan irradiating an important shipping lane and rendering it off-limits.[25]

America, and the world, is further affected when a permanent blackout settles over the Eastern seaboard from Quebec to the Carolinas throwing the global financial system into chaos.[26] The crashing world is easy to imagine partly because so many of its disasters have real world counterparts that are all too familiar, but also because the strain that so many disasters would put on local and global efforts to respond would ultimately undermine those efforts leaving much of the destruction unaddressed. Each disaster zone would become a festering socio-political sore that would continue to erode the stability of an interconnected and interdependent "global village."

A Safe Path Through the Chaos

Throughout the world of *The Massive* glacial melt and rising sea levels have disrupted governments, trade and commerce, communications, and agriculture. Raging storms, earthquakes, and other environmental phenomenon are presented as part of the timeline leading up to the present-day context of the graphic novel. Yet, even with such an intense environmental focus, the environment itself, and the environmental destruction resulting from so many simultaneous problems, Wood's world is not depicted consistently throughout. Ultimately, the Crash world in pictures does not always match the description of the Crash world. When disasters, or their aftermath, are discussed the depictions of them are satisfyingly cheerless and work to generate

6. The Massive *and Life on a Warming Planet*

the "Oh God!" knee jerk response one would expect. However, the conditions rarely match the words as Ninth Wave moves around the globe. Which is not to say that Wood has failed to the same extent that Kirkman or Vaughn do, but in a story crafted to be a warning, the expectations are largely unmet.

Consider the most obvious example, the oceans, which are not only ever-present by virtue of Ninth Wave living on a boat, but also because oceanic and marine conservation efforts are their primary mission focus both pre- and post–Crash. So, beginning with oceans that are already befouled and in dire need of incredible conservation efforts pre–Crash, Wood adds raging oil platform fires all over the world, huge numbers of dead marine life, innumerable shipwrecks, which have already been analyzed earlier so there is every reason to expect that given the storms, terrorism, seismic activity, etc., there would be a lot of shipwrecks *out at sea* and not just in and around harbors. Many of the captains would be the victims of natural events but many others would go down with ships that were victimized as the result of war, piracy, and, in some cases, human ignorance. The oceans, by all reckonings, should be a wreck and, ignoring that some of the Ninth Wave members still swim in the ocean, it is possible that the ocean, despite being described as a liquid wasteland multiple times over, is still large enough that many of the Crash's affects have long since been swallowed up. But consider two examples that present a problem for this idea as the world is depicted by Wood.

One of the biggest concerns the crew of Ninth Wave faces is securing resources—especially fuel—for life at sea. It is a pretty charmed life, all things considered, and especially when considering what the life of the average landlubber must be like during and after the Crash. It is a point underscored when, in Mogadishu, Callum encounters an undesirable acquaintance from his past who claims that, in the post–Crash world, Callum is "truly rich because of [his] ship."[27] But going to Mogadishu is necessary to establish a trade connection with the warlord in charge of what has become the largest black market in East Africa. It is not his only contact for resources, but his time in Mogadishu is a close-up look at daily life for people post–Crash that really is not shown elsewhere with any detail in the early storylines. In addition to Mogadishu, Callum also tries to establish a link with Moksha, a multi-national group of refugees from all around the Arabian Sea that have taken over a large oil platform and declared themselves a sovereign nation.

Moksha is situated on the brink of the Ceylon Abyssal Plain and in international waters, presumably, somewhere around 200 nautical miles south Sri Lanka. The Moksha station was formed four months into the crash and was subsequently joined over the following months by other platform rigs until the "rig nation" had multiplied in size several times over.[28] During Callum's visit to establish a trade relationship with Moksha, Sumon, the leader of Moksha station, recounts the details of the Crash that led to the formation of the Moksha station. The picture he paints is rather grim and Wood's artwork bears that out.

The flood waters of Goa, India are armpit deep to a grown man, Karachi, Pakistan is wracked with political strife as people riot in the streets, refugees flee unlivable conditions along Route 8 in Myanmar, and, most importantly, there is a scene where two very large ships, oil tankers, are burning at sea. Sumon says, "the big oil ships, the U.L.C.C.'s and the F.P.S.O.'s were destroyed to block the Straits of Hormuz and the Aden Gulf. The oil economy was crippled and since then the region has seen nothing but war.... Oil fouled the beaches from Djibouti to Kerala ... the water was so polluted we retched from breathing its vapors."[29]

That is a curious bit of storytelling about this area of the world. Wood has already claimed that the Suez Canal has gone dry, which means the Mediterranean would be spared form sharing in the destruction of the Arabian sea, but it also means the befouling of the sea would be more concentrated. Sumon claims that the beaches are ruined from Djibouti to Kerala—that is, from the northern shore of the Horn of Africa to the western shores of the southern tip of India, just north of Sri Lanka and the Moksha station. If people, whatever their reasons, had done something so calamitous as use ULCCs and FPSOs to block the Strait of Hormuz, separating the Persian Gulf from the Gulf of Oman, at its narrowest point between the United Arab Emirates and Iran's southern shores, *and* to block off the Aden gulf, separating the Red Sea from the Aden Gulf and the Arabian Sea, at its narrowest point between Djibouti and the western shores of Yemen, then the devastation would exist on a colossal scale, as Sumon says.

Adding all the other misfortunes that would be compounding the problems created by the oil slicks and it is reasonable to assume, following Sumon's explanation, that the western shores of India, the entire coastline of Pakistan, most of Iran's southern shores, the entire coastlines of Oman, Yemen, and Djibouti would be completely contaminated, not to mention any island still inhabited in the Arabian Sea or, really, still above water. However, saying the shorelines would be contaminated is misleading because the contamination would actually be at the water's edge, which has already moved pretty far inland all over the world as the seas have risen. Taking these conditions to be true, as Wood represents them, there is a definite problem with Mogadishu.

Somalia's coastline is the western boundary of the Arabian Sea, just a little farther south than Kerala is in the east. In fact, Somalia and Djibouti, just to the north, are geographical next-door neighbors in the Horn of Africa. So, if Djibouti's beaches are destroyed, and Kerala's are destroyed, there is every reason to believe that much of Somalia's beaches are suffering the same fate as the other countries in the Arabian Sea. Yet, when Callum arrives in Mogadishu the oceans are remarkably clean and reminiscent of the pre–Crash and when Mary, one of his crew, *goes for a swim*, the only concern her fellow crewmates have is about sharks.[30]

Mogadishu is partially underwater, but the city is remarkably clean for a thriving third world black market, but more importantly, none of the effects Sumon describes are visible. There are no sick people, there is no oil or evidence of other detritus at the water's edge, and no one choking on the vapors of the oil covered, chemical cocktail, the oceans have supposedly become. There are, however, people busily going about their lives at market, attending church, having tea, and successfully engaging in business of all sorts.[31] In a macabre twist, Callum's undesirable associate makes a point that the two of them should go into the private security business because, post–Crash, it is thriving, but to illustrate his point he indicates the bullet holes still visible from the 1993 Battle of Mogadishu rather than the conditions created by the Crash.[32]

The environmental problems created by the Crash are presented as being a serious, and growing, concern, but the environment largely remains something no worse than a variable the survivors must adjust to before carrying on as usual. Just like Kirkman, Vaughn, and Lemire, the depiction of the world is inconsistent with the description of the world—even when the description of the world is disturbingly accurate. This does a disservice to the actual conditions that the post-apocalyptic would usher in and creates a soft spot in dysto-apocalyptic tales that makes it a desirable future outcome given our present states of affairs. Not only would the fall of governments and global commerce spell doom for certain places very quickly, what agricultural land was still available after rising seas moved inland would be severely put at risk by the in-land migrations of coastal peoples all over the world. Dwindling agricultural land, along with disrupted methods for delivering what little aid could be generated for the hardest hit locales, would make starvation, sickness, and conflict far worse than it seems, even in a graphic novel that routinely mentions near-perpetual war as an outcome of the Crash.

We Are All One and the Same

Conflicts, be they conflicts for control of an area, as a rupture of the trust between citizens and governments, or resulting from pre-apocalyptic identity politics, would not only increase the need for humanitarian aid, but would make delivering that aid even more difficult. The resulting conditions for many of the worlds citizens would be terminal and, as people struggled against their inability to meet even basic needs would, likely, engage in conflicts of their own making to attempt to lay claim to some shred of human decency that would allow them to die in less horrible ways, if not to survive above the bare minimum of basic needs.

Wood routinely mentions the on-going nature of the conflicts, many of them in the early days of the Crash, but the conflicts are never really unpacked beyond a few passing remarks and the occasional frame to emphasize the point. There is one instance where the crew confronts Callum because they desire to return to their homelands to help fight in the struggles faced by people all over the world where companies are still trying to assert their right to make money. Callum describes these conflicts—as so many real conflicts are—as the pointless drawing and redrawing of borders in blood.[33]

Conflicts in a post-apocalyptic world would be fought with increasing brutality as the means of waging industrial warfare became increasingly hard to come by. Fuel shortages would quickly neutralize many aircraft as the availability of, ability to produce, and raw materials for, jet aviation fuel were eliminated by crash conditions. Reserves would be costly, but quickly gobbled up by rich countries and multi-nationals with interests to protect, and, even if it took several months, it would eventually happen because flying would be a luxury few could afford. This point is underscored by Ninth Wave's pilot who makes clear that their helicopter should be used sparingly because they will, according to Mag, Callum's second in command, "never find the avgas fuel those engines need on the black market."[34]

Tanks, large warships, and other military vehicles would be sidelined as well as the resources available to people would be better used in other ways. Unfortunately, scarcity would force some entities to make the decision to use resources for war which would force others to do the same. And, the rate at which certain resources—fuel, bullets, medical supplies, and food— would be consumed would exacerbate problems in non-conflict zones that needed humanitarian aid. Sadly, it is almost a certainty that many people in a post-apocalyptic world would find the use of resources for war to be a good use of those resources. The sequestration of resources by militarized groups powerful enough to collect them would exacerbate problems in non-conflict zones as well making bad conditions worse leading, potentially, to the untimely deaths of thousands.

Throughout *The Massive* conflicts are mentioned as a reality of the post-Crash for a variety of reasons some of which, arguably, could not be avoided, but many of which would continue to be the indefensible warmongering of power hungry elites seeking to solidify their hold on power. In Africa, a continent already ravaged by colonialism and imperialism, and further decimated by proxy wars, wars, and tribalism, Wood describes what conditions there would devolve into as foreign aid dried up and potable water became scarce making five liters of clean water incentive enough to level a village.[35]

Given that Wood has already allowed that the Suez Canal has dried up there is no reason to find fault with his postulation that rampant desertification is underway in areas where weather patterns have so shifted that normally low amounts of rain would become no amount of rain. And, to make matters worse in an already volatile region, it is not difficult to imagine that the entire Northern and Sub-Saharan areas of Africa would suffer immensely, and it would become harder, if not impossible, to live in an already severe place.[36]

What Is Likely Going to Kill Us Isn't Something a Handgun Can Fix

Of course, Africa's deserts would not be the only deserts where conditions might worsen. The North and Central American deserts, the entire Australian interior, as well as the Arabian Peninsula and the Thar Desert in northwest India, could all suffer similar ill-effects. If cool deserts around the world were to undergo similar desertification, just in a cooling direction rather than a warming one, the Colorado Plateau and the Great Basin region of North America as well as much of central Asia could likely become frigidly uninhabitable. Worse, these areas would become difficult to traverse further hampering the movement of goods and services to people in need.

Combined these areas are currently home to millions of people, many of whom would likely die as a direct or indirect result of the disaster events of the Crash, and all the survivors would likely become refugees as the land became inhospitable. Add those numbers to the island inhabitants and coastal populations that would be fleeing rising seas, and even without the conflicts taking place, the world would experience the single largest refugee crisis in history at a time when the global community was least capable of handling such a situation.[37] Unfortunately, Wood never fully develops the tragedy that would be the refugee crisis that would result from a combination of factors such as those that comprise the Crash.

The interruption of trade and commerce, and the swelling numbers of people increasingly forced into smaller and smaller geographical areas would drive the rate of inflation of currency forcing people to use alternative means of determining trade values. Of course, as paper money became increasingly useless, people would turn to the obvious things like gasoline, medical supplies, food, or bullets as ready currency early on, and in most places, which Ninth Wave discovers the hard way in Hong Kong. In Hong Kong, the group is seeking to secure fuel, water, and food and an anonymous dock worker asks about payment. Callum indicates they are able to pay in dollars, kronur, or euros and, Mary adds, they will pay five times the going rate. The dock worker dryly responds that yes, they will pay a lot of cash because "we'll all be wiping our arses with paper money for all [it's] worth."[38]

The point, however, is unsubtly emphasized by Wood while Callum is in Mogadishu in search of resources. The Somalian warlord Callum is dealing with asks how Ninth Wave intends to pay for the goods they acquire and Callum says, again, euros and dollars, to which the warlord responds, "you will get a much better rate when you have hard goods for trade."[39] More than a year into the Crash, however, Wood has the government of Saudi Arabia selling its deep earth water reserves to wealthy clients though what the water was purchased with goes unsaid. Yet, the delivery of the water requires security and the women recruited to the detail are paid with a combination of rations and 15 riyals a day—with no explanation as to why, long after paper money has clearly fallen into disfavor, these women would readily except paper money as payment.[40] The inconsistency here is not fatal to Wood's apocalyptic story, but the inconsistent representation of the African experience during and post–Crash contributes to the overall subtext that the post-apocalyptic world maintains some semblance of pre-apocalyptic normalcy.

More nefarious forms of currency than bullets and such are discussed only once throughout the story, however, and though it is a gruesome example, it provides only a modest inkling of what might be used for currency in the post-apocalyptic as the realities of alternative currencies are a seriously underwhelming aspect of Wood's story. Again, in Mogadishu, as Callum is leaving the city he crosses paths with an old nemesis, Arkady, who is harvesting shark fins to be used as currency. Wood's narrator intones that a bowl of shark fin soup in Hong Kong would sell for more than one hundred pounds sterling.[41] Arkady, for his part, reinforces the post–Crash value of the fins by offering Callum one "one the house" telling him that it "might as well be plated in gold."[42] Obviously, shark fins, as a delicacy, carry with them a lot of environmental concerns, especially sustainability concerns, as well as animal rights and welfare concerns, but shark fins are only one of a host of items that would likely become post-apocalyptic currencies.

The value of certain items in the post-apocalyptic would certainly increase the farther in time society got from the apocalyptic events, but other items, which already hold a high value, especially goods that derive their value from black markets, would not only find wider market shares in terms of value, but would likely thrive as industries in an increasingly lawless world. The on-going collection of items with superstitious or cultural values attached to them, like, say, tiger penis, would have immediate devastating effects on animals and environments. Ivory, pelts, gemstones, like conflict diamonds today, and necessities like salt, would become more valuable than government backed currencies. Given the way that Wood has described the world it is shocking that at no point does the issue of salt arise. If the oceans were as damaged as Wood makes them out to be, then salt would become incredibly valuable as a necessity, at least as important as potable water, not to mention other agricultural items like coffee or chocolate that would become rarer in apocalyptic conditions.

There is one cultural practice that would likely flourish in the post-apocalyptic that Wood never deals with, and it is certainly linked to economic trade—slavery. In 2013, Max Fisher, reporting for *The Washington Post*, using a recently published comprehensive study of modern slavery, estimated that there were 30 million slaves owned worldwide. The highest instances of slave ownership taking place in the very areas of the world that Wood postulates would be the most beset by the events of the Crash.[43]

Unlike Wood, Lemire includes a narrative arc that partially addresses this issue, though he uses it to motivate a longer storyline less concerned with the sexual exploitation of people in the post-apocalyptic and more focused on the development of a deeper main character, he is the only one that captures this reality.[44] However, while many modern slaves are illegally exploited sex workers, many others are not. Anti-slavery advocacy organization Freedom United puts the number of slaves worldwide at 45.8 million people.[45] Not only would the conditions of the post–Crash make it harder for people to resist the pressure to work as indentured servants, sell their children, or their own bodies, but the lack of governmental enforcement of anti-slavery laws would embolden warlords, wealthy elites, and other powerful people to prey upon the weak, poor, or otherwise vulnerable populations. Ultimately, the post–Apocalyptic, regardless of its cause, would see humans bought, sold, and used as currency.

How to Be an Environmentalist at the Not-Quite-Yet End of the World

There are, in *The Massive*, many opportunities to see the world rendered visually as it might look if global warming continues apace. But, as I argued above, the story is not about the environment even as the environment is the most central aspect of the story. *The Massive* does a better job of providing a holistic picture of the environment collapsing from anthropogenic causes than *The Walking Dead* does envisioning a world left unattended in the aftermath of a sweeping viral infection or *Snowpiercer* does with a nuclear winter, but even so, the environment in *The Massive* is not given its full due as the inhospitable and hostile place it would become post-global warming apocalypse. This, of course, carries over into the social, political, and cultural aspects of the post-apocalyptic world, too.

Many of the problems that would arise in the post-viral destruction or nuclear destruction of the world would exist in a post-climate shift world as well, but as Wood has presented it the incremental destruction of the world ought to allow for many of the worst things lurking in the status quo to be sufficiently managed in as safe a way as possible. But if we take seriously the conditions of the world that would exist in the post-apocalyptic, then it seems likely that even with time to prepare there would be a lot of problems from the status quo that would follow us into the post-apocalyptic.

One could argue that there would be lots of people still around to maintain nuclear facilities and keep an eye on things, but that defense falls victim to the idea that the environment can be compartmentalized. As the people move away from the coastal areas overcrowding would breed crime, and disease, and war for limited resources. But the people would not leave all at once, some would see the proverbial writing on the wall before others, and still others would stubbornly stay put until it was too late, the result being that it is completely feasible that even though there are people alive who could, or ought, to tend to certain potentially catastrophic possibilities, there is no guarantee that they will. Regardless, the dead would still need tending to and the living would still need to eat, find shelter, and defend themselves, and in the single-minded pursuit of the necessities it would not take long for minor problems popping up in the aftermath of the collapse to accelerate if we take the potential consequences of global climate shift seriously. In the ongoing saga of a world slowly collapsing around us the instability that certain events would introduce into the norms of society would make certain infrastructure components unsafe—such as high-rise buildings—and others radically dangerous—like dams breaking.

Dams are a serious problem, especially in industrialized countries, because we have been building them literally with reckless abandon. There are more than 57,000 large dams worldwide and more than 300 hundred major dams—the majority of both exist in China and the United States.[46] If a dam were to be compromised as a result of Crash events, through mismanagement or lack of maintenance, seismic activity, or, worse, through warfare or sabotage on par with Edward Abbey's Monkey Wrench Gang, downstream devastation would be phenomenal by itself, but as a contribution to the matrix of apocalyptic occurrences, each dam failure would be a reckoning in its own right. Consider two examples, one fictional and one real, to give context to this concern that neither Wood, nor anyone else analyzed so far, addresses.

If the Hoover Dam was compromised, it would release the more than 10 trillion gallons of water currently held in its reservoir into the Colorado River valley. That is enough water to cover an area the size of Connecticut in 10 feet of water.[47] If the Hoover Dam's failure were the result of the Glen Canyon Dam failing upstream, their combined water would represent approximately four years' worth of normal river flow occurring at once.[48] The tidal wave of water from the dam break would move rapidly in a tsunami-like fashion which would sweep away population centers along the river corridor including overwhelming downstream dams—the Davis dam and Parker dam—incorporating their water into the fury of the Lake Mead onslaught—these would include Lake Mohave's 500 billion gallons of water and Lake Havasu's 200 billion gallons of water. Not only would this result in innumerable deaths, it would compromise the commercial water delivery of an area from central Arizona, through southern Nevada, to the Pacific Ocean and the devastation would stretch into northern Mexico.

Compare this to the Banqiao Dam disaster in China in 1975—a disaster whose details only began to emerge in 2005 thanks to the release of governments documents. On August 8, fighting a losing battle against a raging storm and a rapidly rising Ru River, the residents of Henan Province experienced one of the worst catastrophes in history—an experience that was quickly covered up by the state. At approximately one in the morning the Banqiao Dam broke sending 184.8 trillion gallons of water cascading through the countryside.[49] That amount of water was capable of sweeping away whole towns and villages and killing in excess of 170,000 people. Like the United States in the past, China is set to go on a dam building spree in the coming years, ignoring the potential harms of doing so—increased landslides and droughts—even though dams provide little by way of economic gains and are known to be environmentally harmful in numerous ways. But, in an apocalyptic scenario, where synergistic events cause disruptions globally, alerting people to potential dam failures, to say nothing of calls for help, would be difficult once communications broke down and, such failings would only exacerbate the problems.

Most post-apocalyptic stories make a lot out of failing or failed communications systems, and the lack of reliable communication, especially in a world as interconnected as ours is now, and one that was being ravaged by natural events of epic proportions, would be a major disadvantage. Wood's survivors are largely able to keep their communication systems operational, not to mention that, though there are shortages, few people anywhere actually seem to be put into debilitating scenarios by those shortages. These breakdowns in communication, resource acquisition and dispersal, governance, coupled with increasingly harmful environmental situations, to say nothing of the devolving socio-political day-to-day existences of so many people, would leave many at-risk populations more vulnerable than others. It is a little surprising to find no mention of the hazards faced by the disabled populations around the world represented in Wood's story, nor the elderly, or sick. The presumption is that certain populations would face a rapid die-off in post-apocalyptic conditions, something that prose writers are more inclined to tackle in post-apocalyptic literature, but perhaps, like the violence of sexual slavery, it is too much to depict the horror of leaving physically or mentally disabled individuals, who are otherwise healthy, to die as we flee for our own lives one step ahead of disaster.

If *The Massive* is meant to be a morality play to give us the opportunity to answer for ourselves Callum's toughest question—"What does it mean to be an environmentalist at the end of the world?"—then the authors and artists of graphic novels have an obligation to render the environment accurately, but also to make that question central to the overall story.[50] Though Wood gives us a robust engagement with the ideas critical to beginning to answer that question, the answer remains, I argue, muddled. The story is hopeful, not just at the end, when the world is reborn, but throughout, as if grappling with the most difficult questions facing us requires that our realism make room for the notion that life will go on, but not just that, rather something like *a life worth living* will go on, and that is, now and in the post-apocalyptic, highly unlikely unless we start to come to terms with what it means to be an environmentalist *now*, an environmentalist not-quite-at the end of the world.

7

Environmental Theory in an Apocalyptic Age

We Lack a History of the Future Tense

The preceding has been an attempt to get beyond a brief sketch of the social and environmental problems that would likely manifest in the event that human civilization falls victim to an apocalyptic event. However, even if the brevity of the accounts or inaccuracies contained therein miss the analytic mark, it is certain that the environment must take a more central position in our use of both utopias and dystopias, apocalyptic or not, as critical devices. This is especially true if we are going to see the social critiques presented through such imaginative devices motivate the kinds of human behaviors and social revisions necessary to avert the catastrophic outcomes that could manifest in the future if a significant apocalyptic disaster should occur. The average citizen does not, unfortunately, read much critical theory; so, it is even more important that both fictionalized and documentary accounts of the future accurately depict the future that is *most likely to develop out of our current circumstances*. Of the utmost concern here is the fact that for most people their relationship to nature is one of alienation—a comfortable separateness that allows them to think of themselves as detached from the destruction of the natural world. But, as Walter Benjamin notes, this is a type of alienation that has reached a point where humans can now "experience their destruction as an aesthetic pleasure of the first order."[1]

Given how we have built the world around us life in the aftermath of a global apocalyptic catastrophe would neither last long nor be even moderately comfortable in the conditions that would await us and all other forms of life that managed to weather the apocalyptic event. Hence, the solutions to such concerns are not to be found in the bravado aimed at overcoming those conditions and refashioning a new and better society on a *tabula rasa*, but rather in re-thinking and re-fashioning society now—how best to accomplish that is likely a story best told by utopian rather than dystopian storytellers.

It is also certain that the narrow focus and brevity of my arguments has left noticeable holes in the larger, more complete, ecocritical approach to dystopia and apocalypse that would merit a much longer, more thorough, and likely multi-volume discussion to accomplish. Graphic novels are but one form of artwork that broach the subject of apocalyptica, and a fairly new one at that as far as mainstream popularity goes, but they do fit more comfortably into a society beholden to imagery and, if our social impulse is a narcissistic one, then few art forms have a comparable ability to hold up a mirror for us to gaze into. It may seem that drawing out the failures of graphic novels to properly represent the post-apocalyptic world is a waste of time, after all, if the world is on the brink is there not a better way, a more meaningful way, of making these arguments than examining comic books?

I do not think so because art may well be our best bet for re-framing current environmental discussions. More importantly, graphic novels are in a prime position to deconstruct the realities of the present, to breakdown what has become in other art mediums, as Timothy Morton labels it, the aesthetic as anesthetic.[2] Apocalyptic graphic novels generally feature characters who occupy a duplicitous nature: on the one hand they are, usually, common folk that are easily identified with and, on the other hand, they hold the hoped for position of the messianic figure—the person that will survive the apocalypse and bring about a new world. These figures gather up collective desires into stable objects of hope, serve as anchoring points for dreams, and produce the conditions for collective action. As messianic figures they embody expectations of sweeping change, salvation from apocalyptic destruction, and radical transformations of societal norms. More importantly, unlike their real messianic counterparts, the characters that navigate the apocalyptic and save humanity never disappoint.[3]

Perhaps some will counter that there are plenty of post-apocalyptic stories that present a landscape that is forbidding on all levels. However, the counterclaim to such an assertion is that while some do portray a terrible world to inhabit, the landscape itself is usually presented as completely destroyed, a mad max-esque wasteland largely unaddressed in desolate monotone uniformity with a lack of specificity, perhaps to allow the reader to imagine the horrors that might await in such a landscape.[4] Even drawing this comparison between the post-apocalyptic and Mad Max is done by theorists that want to contextualize the difficulty of inhabiting the world in an apocalyptic aftermath. For instance, Bill McKibben, makes this comparison arguing,

> If you think about the cramped future long enough, for instance, you can end up convinced you'll be standing over your vegetable patch with your shotgun, warding off the marauding gang that's after your carrots…. The Marines aren't going to be much help there—*they're not geared for Mad Max*—but your neighbors might be. Imagining local

life in a difficult world means imagining taking more responsibility not only for your food but *for your defense*.[5]

Christian Parenti, responding to this position, argues that this presents America as a failed state and he would be correct, but the problem really is that both McKibben and Parenti are mistaken.[6] McKibben's mistake is thinking that militias will be able to defend themselves and their vegetables patches, but ignores the fact that most people will not be in a position to be stationary in the post-apocalyptic aftermath—he is not thinking holistically enough about the reality of post-apocalyptic life. For Parenti, the mistake is one token and not type, because he is correct to identify the United States as a failed state if these conditions prevailed, but *how* the failed state would function is a more nuanced and difficult question to assess. This is, by and large, a problem that Kirkman addresses, albeit inadequately, but nevertheless gets more right than McKibben. Parenti, conversely, is correct in asserting the image is one of a failed state, as most states would be, as correctly envisaged by Wood's storytelling in *The Massive*, but Parenti fails to conceptualize what that means. He goes on to argue that there ought to be a better way to put the wealth and resources generated by capitalism to better use than continuing or exacerbating existing societal ills. This argument only finds real purchase in a pre-apocalyptic world. Can we not, right now, figure out how to better use the wealth and resources at our disposal to *avoid* an apocalyptic fate? If we cannot, then it is a moot point to wonder if those resources might be better used after the fact because, after the fact, there will be no infrastructure for doing so and the mad scramble to secure what resources will exist will determine the nature of the conflict for those resources.

My contention throughout this text is that most people tend to easily imagine a horrible post-apocalyptic landscape, but again, one which regardless of its attendant horrors can be overcome and reconquered, and, after it has been conquered, life can resume its normal and familiar rhythms, society can begin anew. In this post-post-apocalypse life on earth will begin again with a renewed vigor and earnest convictions that a utopia can be fashioned on top of the ruins. All the while ignoring that the apocalyptic landscape somehow manages to provide for the needs of the lucky few while they re-establish themselves as the dominant species on earth. There will be little opportunity for putting down roots—social or vegetal—for quite some time in the aftermath of an apocalyptic global collapse.

The perpetuation of this misconception, even among the best portrayals of the post-apocalypse, is why a synergistic understanding is desperately needed in our social imaginings and why the environment must take center stage in the dystopian narratives of the post-apocalyptic—if for no other reason than to undermine the current hopefulness such stories seem to breed. If there is going to be adequate and honest engagement with environmental concerns, then there must be an idea of the environment that is more than just the vague and ephemeral greenery filling in the background. In order to have an environment there must be a space for it and in order to have an idea of an environment there must be *ideas of space and place* that will allow for a clearer picture of what the environment *actually* is in the first place.[7]

The space and place for environmental understanding is not the wilderness, though Romantics would have you believe that getting lost in the woods is the best way to find yourself, coming out of the woods is the best way to find your place in nature. Humans have lived in nature for the entirety of their existence, but recently anthropocentric and speciesist conceptions of humans relative to the rest of the world have separated humans from nature and, equally important, their nature. Marx would call this a form of alienation. Living communally in cities designed to cater to human needs and subjugate the natural world has fostered and exacerbated this experience of alienation so much so that most people no longer think of their day-to-day lived experiences taking place *in nature*. But if that is even remotely true, then re-thinking the city as a dystopian staging ground ought to prove beneficial for understanding how to re-frame our environmental theories with respect to avoiding the worst-case apocalyptic scenarios.

Civilization Is in Crisis

It would not make sense to think of bees without at some point addressing hives, nor birds without considering nests, so if the focal point of apocalyptic ecology is thinking about the human in terms of the environment, then it only makes sense to use the cityscape as the point of access to understanding how modern environmentalism ought to move forward given the circumstances in which we inhabit the earth. Timothy Morton rightly claims that the idea of the environment is more or less a way of considering groups and collectives—humans surrounded by nature, or in continuity with other beings such as animals and plants; it is about *being-with*.[8] I think this is on the right track although I have challenged this idea and supplanted it with a strong notion of Being-as-Belonging.[9] Along the same interpretive lines that Morton utilizes to craft his critique of nature in relation to a more sensitive ecological awareness, Warwick Fox argues that environmental ethics has become a misnomer in modern usages. Fox argues that the problem inherent in the idea of the environment is that when most people, especially those in the developed, industrial West, look around the world they see people, other animals, trees and plants, rain clouds, and so on, but they also see buildings, roads, cars, and other components of city life.

Our "environment" consists not only of a self-organizing, natural environment but also of an intentionally organized, artificial, built, and otherwise human constructed environment. Hence, when we speak of the environment we tend to mean the undeveloped green space outside city limits—the natural environment—and we do not think to include where we live. Environmental ethics, then, finds itself at odds with other social ethics in attempting to determine the best way to inhabit the environment. Fox advances a reconceptualization of our relationship to nature along the lines of a responsive cohesion that integrates more fully human experience and human existence in a totalized environmental setting.[10] Here it may prove helpful to extend Morton's arguments, and my own, to capture the cityscape as the location of what Eben Kirksey calls "ensembles of selves." If we conceive of human communities as ensembles of selves we must mean all of the associations which compose the relations of reciprocity and accountability between conscious agents who are entangled with each other.[11] This requires us to shift our thinking about the city so that we understand that space as more than where humans live and instead as a place where lots of beings live. Our cities should reflect our empathy and concern for humans as beings that belong to a larger biotic community which also lives in cities with us.

This idea prompts another question about the conflation of environmental ethics with environmental philosophy. We might be tempted to think that if we were to figure out how to live more *ethically* with respect to nature, that we would then not need to worry about environmental questions *generally*. The assumption here, the one that Fox and others are working to dispel, is that the environmental crisis is *merely* one of values and ethics. It is not. In fact, the genuine philosophical issues that underlie environmental ethics are deeper and richer philosophical problems. It is definitely a problem that our current attitudes and actions are destructive of the environment, but, even if we managed to correct that and globally achieved sustainable living, we would still need environmental philosophy to ask, understand, and answer questions about the fundamental nature of human-environment relations. Kevin Laplante argues along these lines that by

> attempting to answer the standard questions of environmental ethics and radical environmental philosophy, environmental philosophers invariably run into questions concerning the ecological dimensions of human perception, cognition, and activity … the core questions for environmental philosophy…. Contemporary environmental philosophy is what you get when the legitimate fears and concerns of environmentalists are brought to bear on philosophical thinking about human-environment relations. The standard problems of environmental ethics and radical environmental philosophy are a natural outgrowth of these concerns.[12]

The Rhythm of Knowing

Generally speaking, our conceptions of dysto-apocalypses begin in, or have their roots in, some origin beyond our control that happens suddenly and involves technological devastation, massive fauna and flora population loss and leaves the world largely inhospitable, usually a cataclysm that sunders urban, rural, and wild life—and that is how I have treated the apocalypse throughout my analysis. Rarely do we think of dystopia as incremental, that is, as something that festers in our cities and oozes its way upward and outward, spreading like a disease. Dystopia is usually conceived as what happens after a society is brought to the brink of a major life-as-we-know-it event or some colossal tragedy occurs and in its aftermath the few remaining survivors struggle in a disorienting new world or they are organized into a state organism which controls their every move ostensibly "for their own good."

In theorizing utopia and dystopia that tends to be the focus, "our own good," what that means, how to achieve it, and how to recognize it when we have. Theory of this type is not all fantasy though it does tend to pop up in fantasy more regularly than elsewhere. Dystopias in the wasteland category usually result from our having too much of our own good whereas the tyrannical sort often bear the unmistakable hallmark of having been, at one time, attempts to provide "the good" in equal measure to people. But dystopias and apocalypse must happen *somewhere* and by placing our focus on the global instead of the local, by imagining the apocalyptic instead of the dystopian, it is possible to reconfigure an ethical outlook regarding not only where we live but how we live. A must if we are to avoid the pitfalls of ecological apocalypticism as it currently exists in pop culture cultivating a death wish by trying to see beyond death.[13]

If, however, we take a step backward and examine our dystopian narratives from a broader perspective we would see that dystopia creeps up on us until, before we know it, we are consumed by it. These little bits of incremental dystopia have certain recognizable features: they are generally and widely mistaken for positives, they are widespread, and they actively contribute to the compounding of other negative elements of human social existence normally associated with dystopian futures. Urban green space is just such an incremental dystopian aspect so social life. Urban green spaces are disassociated with nature because they are often well maintained and manicured, that is, they lack a wildness quality often associated with nature. Urban green space is the locus of alienation and displacement, but because urban green space is unevenly present in cities the alienation experienced by people worsens as the green space dwindles. Thus, alienation of this sort applies asymmetrically to the lower classes and inner-city populations much worse than their upper class and suburban counterparts.

To fully address this issue, I want to provide a two-pronged analysis. On the one hand, I want to give a brief account of the three criteria I have associated with incremental dystopia and on the other hand give an accounting of how green space functions as dystopian space in an urban environment. In

7. Environmental Theory in an Apocalyptic Age   149

Spaces of Hope David Harvey argues that utopian thinking has been the fuel that drives the engine of urban planning.[14] We are often tempted to think of the environs of the urban setting as fundamental to a healthy and equitable social life, but too often, especially in modern and contemporary approaches, we divorce urban environs from the environment.

This is a problem more generally because it is far too common to conceive of ourselves as "apart from" rather than "a part of" the environment. The result is that our city planning and urban development often allow for green "spaces" that are not capable of providing to human social existence what a healthy and incorporated environment would and does. What we have created is an archipelago of little faux utopian islands surrounded by a vast sea of dystopian landscape that is, by degrees, worse in some places than others. However, my position is that these islands of green are not only not utopian outside of their theoretical construction, but are, in fact, dystopian spaces. Garret Eckbo captures this nuance nicely.

> Today what architects and planners call the built environment is an apt term separating the urbanised world from what is left of the natural world, the agrarian world, and the resource exploitive world from which come our coal, timber, oil, and minerals. Although the urbanised area of the United States covers only about ten per cent of its total area, the other ninety per cent serves that built environment as a resource base and recreation area. By a strange and complex process the urban centres, shining stars of human culture, have become totally parasitic on the balance of the land, milking it of resources with vast energy and ingenuity, and returning to it only waste, asphalt, pollution, and destruction.[15]

Green spaces are often taken to be acknowledgments of our dependence upon nature and its necessity as a function of a healthy urban existence. Green spaces included in urban planning, however, capture our dependence upon rather than our co-existence and interdependence with the natural environment. In post–World War II societies, the scramble to sprawl was precipitated by the upwardly mobile middle classes as the aspiring suburbanite sought to possess a little breathing room in the form of a small yard, perhaps with a tree, hemmed by a white picket fence. The clamor for personal space, the rush to vacate population dense areas, in addition to the new mobility of an economically secure middle class, laid the ground work for an individualist subjectivity that obliterated a recognizable dependence on nature and replaced it with a nature understood as the site of recreation and the home of undomesticated animals.

Some modernist planners were keen to try to maintain the contextual need for an integrated environment rather than embrace the increasingly popular notion of the separation of the natural world from the urban(e) one. Le Corbusier's *Radiant City* is an example of the attempts being made to maintain the integrity of urban and natural co-existence. Le Corbusier offers a truly comprehensive conception of urban planning conjoined with radically ambitious projects of political and economic restructuring, endeavoring to promote urban settlements founded upon the principles of social solidarity rather than segregation but the possibilities of incorporation inherent in such programs were never completely fulfilled.[16] What is lost in the division of urban and natural life, when they are cleaved apart and both understood and experienced separately, is a group identity and cultural behavior that is sensitive to and aware of the need for a healthy, protected, and fully integrated natural world. Once the natural world is understood as different, as completely separate, it is much easier to justify its destruction.

In spite of the failure of integrated urban planning to take hold modernist attempts continued to influence urban planning throughout the post–World War II years through contemporary times. Unfortunately, the use of green spaces to define utopian spaces—or, put conversely, the recognition of "green voids" to define dystopian spaces—has contributed to the increasingly insubstantial use of environmental features to placate urban dwellers. What once was the standard for utopia or dystopia has become the general baseline for acceptable minimums in urban planning. As such, the inclusion of green spaces or the integration of green technologies have become the watchword for and standard of utopian spaces even when these spaces are creating the conditions for a dystopian existence.

We Lack the Language to Fully Communicate These Ideas

Featuring the escalating extremes of wealth and poverty so often lamented by past urban planners, "but coupled with an intensified fiscal austerity to meet the rigours of global competition, [the contemporary city] appears to be manifesting as an intensely uneven patchwork of utopian and dystopian spaces that are, to all intents and purposes, physically proximate but institutionally estranged. Put another way, the densely settled and heterogeneous 'worlds' that make cities such vibrant places appear to be premised increasingly upon 'indifferent worlds and detached lifestyles.'"[17] As David Harvey understands the problem, "the effect is to divide up the urban realm into a patchwork quilt of islands of relative affluence struggling to secure themselves in a sea of spreading decay."[18]

At this point I would argue that we have exacerbated the problem beyond what can be accurately captured by the concept of a "patchwork quilt" or, as I argued above, an "archipelago" because decades worth of disengagement with and separation from nature has left our urban landscapes devoid of actual natural spaces of the kind deemed necessary to provide societal benefits.

7. Environmental Theory in an Apocalyptic Age   151

These spaces are inefficient for recycling the waste produced by cities, they produce squalor not only at the lowest and farthest edge of socio-economic status, but scarcity even among animals which are driven further and further into settled areas to find sustenance. What begins as an aboriginal hunter stalking through endless jungles ends, today, not only in the guise of the homeless citizen sleeping in a concrete desert, but also in the form of the addled (sub)urbanite attempting to take advantage of the false respite provided by manicured lawns, overly small heavily designed parks, and unusable tracts of land fenced in and masquerading as "wild" spaces. It is no wonder, given the political climate today, coupled with the social realities facing many people that our vision of the future tends to be blinded by a hope for destruction.

Urban life, and, increasingly, its suburban counterpart are being transformed into spaces of rigorous heterogeneity and consumption. This is particularly the case in spaces of consumption like urban shopping malls and suburban shopping centers where the incorporation of "theme parks, rides and amusements and multi-screen or multiplex movie theatres" has led to a faux diversification in, and proportional intensification of, consumption girded by the self-deluding concept of environmentally friendly consumption.[19] Moreover where developers introduce globally celebrated iconography—such as that associated with the Disney Corporation[20]—the utopian mirage is accomplished inside the deteriorating oasis of society. It is a hollow utopian experience that makes possible the dystopian hope which fuels an appetite for erasure of societal norms. Once the idea that a future apocalypse can be liberating settles into a communal psyche it is much easier to feel like contributing to the destruction of the social order is an amoral choice.

For some theorists, such as Christine Boyer, these centers of spectacle have the powerful capacity to erase the distinctions between the orchestrated spectacle and the engulfing dystopian cityscape. Boyer argues, "the awareness of highways in disrepair, charred and abandoned tenements, the scourge of drugs, the wandering homeless, deteriorating transport networks—all are erased and ignored in the idealized city tableaux set up before the spectator's eyes and presented as an entertaining show."[21] Such an understanding of the condition of urban planning and contemporary cityscapes parallels and highlights the slow obstruction of progressive values as "naturists" seeking to secure preservation and conservation fail to envision "integration" as a necessity for any successful urban plan. What currently passes for "integration" is little more than tree-lined streets, small parks, and green technologies that often shift a problem from one venue to another, albeit often less harmful, venue. A major problem with articulating the problems of environmental apocalypticism is that the post-apocalyptic cannot fictionalize a world with a broken environment as part of its narrative because this is an aspect of many peoples current lived experience which the apocalypse is supposed to erase.

Urban planning and the lack of green spaces, or again, urban planning and the growing acceptance of green voids, is having a direct and indirect effect on flora and fauna as well as ecosystem and watershed health. The result is not just the disruption of an environment, but rather the destruction of environmental health. A destruction that continues unabated as we allow for greater and greater usages of "green spaces" which are increasingly areas identified as undevelopable, non-arable, or marginal. These spaces are too often significantly incapable of maintaining a healthy ecosystem not only because of their insufficient size but also because they are besieged by human activities on all sides. We are left with a fundamental conflict between the dominant exploitive socio-cultural attitude toward nature, expressed in cities that inculcate an overall destructive lifeworld compounded by pollution which affects the health of people, the environment, and even the political and economic systems which generates the problems.

This conflict confronts us with certain alternatives, not necessarily total, but embodying the need for attitudinal and policy-making changes.[22] This also has dire consequences for how we imagine city-spaces not just how we inhabit them. Before grand boulevards led to places of power there was a multiplicity of gatherings around common causes, shaped by the forces of thinking and imagining, that was connected by a labyrinthian matrix of streets. These gatherings were communities but not in isolation from one another and none had the power to determine the others and each demanded to be recognized and respected.[23] Cities now are designed to be efficient and frictionless to produce productivity rather than creativity. Skyscrapers jut upward cleaving imagination and feeling from thinking, but the cold logic of global logistics creates daydreams where the horizon is black smoke, ash, and silence.

A Pair of Deuces Ain't Much

In contemporary cities, then, the social groups constituting a sharply splintering class society appear to be negotiating particular time-geographies, snaking their respective paths along a strict compass of localized and dissociated landscapes, at times giving rise to a tense, often anomic and alienating urban fabric. These architectural, political, and economic contours raise profound questions about 'the city' as an object of analysis and also about the future expression of citizenship, spatial justice, and urban politics.[24] Political ecology develops this idea by taking a broad view of disasters both in kind and scope, but as is too often seen in theoretical schemes dealing with disasters the focus remains limited to a non-global event. Political ecology nevertheless emphasizes that the "conjunction of a human population and

7. Environmental Theory in an Apocalyptic Age 153

a potentially destructive agent" need not produce a disaster event, but "a disaster is made inevitable by the historically produced pattern of vulnerability, evidenced in the location, infrastructure, sociopolitical structures, production patterns, and ideology, that characterizes a society."[25]

Communal fragmentation is expected as survivors struggle to define their existence in relation to what has happened. Their only initial reference point will be their pre-apocalyptic existence which defined their existence along race, class, and gender lines. So, on the one hand, many people will see disaster and reconstruction as an opportunity for major changes in the way things are done while, on the other hand, others will want nothing more than to continue with the status quo, to reestablish a degree of consistency and continuity with the past. The notable difference between the local and the global is not in the evolution of the disaster, but rather in the fact that the latter point of view of returning to the status quo is literally not an option. This pits the survivors against a hostile and foreign environment with no expectation of help. In an apocalyptic world the environment, or what is left of it, will determine how and when the survivors will be able to institute major changes in the way things are done.[26]

The ability to extrapolate conclusions drawn from the localized problems that exist to a global scale that properly renders those consequences realistically is best achieved through an analysis of the family resemblances that exist between the two. Wittgenstein's concept of family resemblances is likely the best means for explaining why the environment needs to take a more pivotal role in the narratives that depict the post-apocalyptic future. In the aftermath of a localized disaster people and groups are faced with a multiplicity of problems that arise during the "often lengthy evolution of a disaster: through warning, impact, emergency, relief, recovery, and reconstruction."[27]

There are certainly important links between a localized disaster—Fukuyama Japan or the tidal tsunami that wrecked India—and a global disaster that fundamentally alters the ability of life to continue under normal circumstances in a multiplicity of places simultaneously. Wittgenstein employs the idea of spinning thread to explain such a wide-ranging phenomenon that is "a complicated network of similarities, overlapping and criss-crossing: sometimes overall similarities, sometimes similarity of detail."[28] An event like a global disaster constitutes a set of family resemblances which require only a minimum of definitional boundaries to fully comprehend.

Pre-apocalyptic urban life is a result of the combination and interaction of earth, water, rock, vegetation and animal life, as well as construction and people. This life is, in many ways and for many reasons, experienced rationally and irrationally, as the triumph of humankind; a congratulatory conception that is only possible by ignoring the pillaging and destruction of nature, as science and art, technologically, religiously, ideologically. Urban life can only be hailed as an evolutionary success by glossing it as a cultural triumph while simultaneously being blinded to the visual disaster it represents, by defining it as an economic gold mine by simply valuing nature only monetarily. The future of the natural world hangs in the balance of political usage—the less votes it can garner the more likely it is to be plundered. Basic to urban living is the long-term drive to conquer nature in order to achieve security.

Our momentum is such that we forget that we have already conquered nature and are beginning to live beyond nature's ability to provide, in a technological Never-Never Land.[29] If we are not careful, we may soon find the strongest family resemblance between the local and the global is that they are both dead. Which, in a funny-not-funny sort of way, is exactly why a story like *The Walking Dead* can generate so much desire and hope among its readership that such a thing will come to pass—everyone thinks they will be a survivor and not a zombie—everyone wants the opportunity to *change the way things are*. This is in all actuality precisely the kind of stupidity that Isabelle Stengers decries as indicative of the "I am aware but all the same..." stance which has replaced thinking, and, in my argument, especially for people that fetishize apocalyptic destruction as an inevitable outcome.[30]

Alas, and Also

In this regard we could rightly question the accuracy of the art employed by graphic novelists to interrogate the possibilities that lurk just beyond the horizon of our present circumstances. Is there a real, that is, true, reason to be genuinely concerned about the yet-to-come apocalypse? Or, conversely, is this art, and, perhaps, art generally, an exercise in entertaining falsehoods—that is, can there be any truth worth taking seriously in fantastical artistic representations of the future, especially the apocalyptic future? J.L. Austin, to the consternation of philosophers and culture critics, argues that the language of art is *parasitic* upon the same language uses uttered *outside of an artistic context*.[31] While Austin fully intended to cordon off the serious scholarship about language from the playfully artistic "it is precisely here—in the realm of the perverse, or the explicitly performative—that things get interesting."[32]

It is in the very playfulness of imagination that we are best able to skirt the issues of language proper and properly used and instead focus on meaning, intent, value, and virtue, ultimately asking ourselves, what is possible, why do I feel this way or that, what should my interpretation or understanding of myself be? Art is a way for us to experience ourselves more deeply and to come to terms with certain truths out in the objective spaces of co-existence—namely our own morality and mortality. Though everyone ostensibly knows they will die we are no longer comfortable thinking about the finiteness of our existence and, the result, is that we are too comfortable ignoring the reality that we are hastening our own demise by killing the natural world.

It should go without saying that as natural things, if we destroy nature, we are, without exception, acting self-destructively—we are literally killing ourselves and our salvific aspirations increasingly lie in apocalyptic hope. I find that this idea resonates with Lauren Berlant's concept of cruel optimism. However, I think that while parallels exist Berlant's concept is aimed, much like Susan Hoffman's, in social crises that are not totalizing even if, in Berlant's case, they are conceived of as bigger than the catastrophes that form the heart of Hoffman's argument. Berlant explains cruel optimism by premising that all attachments are optimistic where optimism is the force that moves you out of yourself and into the world to bring closer the satisfying something that you cannot generate on your own. She stresses, I think rightly, that optimism may not feel optimistic, in fact, it may feel like dread, anxiety, or any of a range of emotive expressions tethered to the prospect of the change that is coming. The important parallel to apocalyptic hope is Berlant's assertion that optimism becomes cruel "when the object that draws your attachment actively impedes the aim that brought you to it initially."[33] I would contend that desiring a transformed, improved-for-the-better, society which can only be achieved through utter destruction is a perfect example of cruelly optimistic apocalyptic hope. Indiscriminate, totalizing, mass death should not be a requirement for motivating individual and societal change.

An unknown author in a late medieval disputation poem ruminates on the necessity of thinking about death before it is too late. Not to stoically come to terms with the reality that we will all eventually perish, but to think of what we are doing with our lives, and, in the spirit of the medieval Christianity which produced the poet, to think about what a good death means and what its value is. Though the sentiment is one that has been ever present in human society, his words still ring true today, and perhaps truer, given the state of the world, but nevertheless, it bears repeating: While your grave is still undug, it is good to think on death.[34] If we are currently living a life of self-destruction and our salvation lies in total destruction, then, perhaps, now more than ever, we need artworks that help us examine why we are living the way we are and what, if anything, we can do to change it. To accomplish that we need artworks and philosophies that honestly reckon with what it means to die a good death, but more importantly, what it means to live a good life.

Chapter Notes

Preface

1. Scott McCloud, *Understanding Comics: The Invisible Art* (New York: HarperCollins, 1994): 3. Both negative depictions of comics are provided by McCloud in his introduction to what a comic is and can be.

2. An excellent example of this is Cormac McCarthy's *The Road* (New York: Vintage Books, 2006) which presents a harsh post-apocalyptic world to the readers but even with the stark poetic rendering of life in the aftermath, replete with the horrors that develop in the fallout of failing social orders, it is difficult to grasp more than the horror with the imagination especially when considered alongside the graphic novel counterparts that attempt to conjure the same experience. This is, to some extent, a drawback based largely on the limitations of language.

3. Hillary Chute, *Outside the Box* (Chicago: University of Chicago Press, 2014): 25.

4. Hillary Chute, *Disaster Drawn* (Cambridge: The Belknap Press of Harvard University Press, 2016): 17.

5. Apocalyptic destruction has to be contextualized as a hyperobject in order to begin to grasp the scope, scale, and devastation singled out be theories dealing with apocalyptic calamities.

6. The possibilities here range from rearranging the way a story is presented in each panel to what is revealed in each panel to hints and clues sprinkled throughout the narrative structure to help "fill in" the story that was omitted from depiction both in the narrative and in the art itself. One of the most important tools available to comic artists in this regard is the gutter which is the space that separates on panel from the next and requires the reader to provide the context for the connection. Scott McCloud dedicates an entire chapter of *Understanding Comics* to the importance of the gutter and its uses. See also Hillary Chute, *Disaster Drawn* (Cambridge: The Belknap Press, 2016).

7. Susanna Hoffman, "Anthropology and the Angry Earth: An Overview," *The Angry Earth*, eds. Anthony Oliver-Smith and Susanna Hoffman (New York: Routledge, 1999): 1.

8. Christopher Ingraham, "The Long Steady Decline of Literary Reading," *The Washington Post* (September 7, 2016), accessed online. Ingraham's article contains links to several relevant studies, including the NEA study cited above. National Endowment for the Arts, "Results from the Annual Arts Basic Survey 2013–2015," published online, available at https://www.arts.gov/artistic-fields/research-analysis/arts-data-profiles/arts-data-profile-10.

9. Andrew Perrin, "Who Doesn't Read Books in America?" Pew Research Center, March 23, 2018, published online, available at http://www.pewresearch.org/fact-tank/2018/03/23/who-doesnt-read-books-in-america/.

10. There is a lot of research on both sides of the divide regarding the effects and extent of those effects on high consumption viewers of electronic media. Much of that research is tentative, and cautious, often amounting to little more than observational hypotheses, but my claim here is more basic. I am merely asserting that someone who regularly engages with electronic media is less likely to be willing to do the work of imagining in conjunction with literature as evidenced by the number of people who identify reading

as "boring." Recently, Naomi Baron captured succinctly what is going on beneath the surface of the "boring" claim in a conversation with a student. Though an avid reader of hardcopy books he explains that physical texts are too static when compared to the distractions offered by electronic media. Naomi Baron, "Is Reading Boring," *Huffington Post* (May 18, 2015) accessed online, https://www.huffingtonpost.com/naomi-s-baron/is-reading-boring_b_6894200.html.

11. Scott McCloud, 59.

12. For more on the use and importance of images see Gwyn Prins and Elizabeth Sellwood, "Global Security Problems and the Challenge to Democratic Process," *Reimagining Political Community*, eds. Daniele Archibugi, David Held, and Martin Kohler (Stanford: Stanford University Press, 1998).

13. Scott McCloud does an excellent job of explaining the relationship between words and images as part of a unified vocabulary. His assertion that words are the ultimate abstraction (47–53) corresponds nicely with my claim that, left to their own devices, many readers are not going to accurately conceive of the post-apocalyptic world. It takes an exceptionally well written post-apocalyptic story to capture what is really at stake in our failing to take seriously the possibility of environmental collapse.

14. For a thorough treatment of comics and graphic novels, both in terms of their historical development as well as their socio-cultural determination, importantly disabusing the idea that a "comic" cannot or should not be taken seriously as academic fare, see *The Graphic Novel*, ed. Stephen Tabachnick (Cambridge: Cambridge University Press, 2017).

15. For a solid analysis of *Watchmen* as a coming-of-age experience both for readers and comics as a genre see Iain Thomson, "Deconstructing the Hero," in *Comics as Philosophy*, ed. Jeff McLaughlin (Jackson: University Press of Mississippi, 2005): 100–129.

16. Husna Huq, "5 Reasons Graphic Novels are the Next Big Thing at Your Library," *Christian Science Monitor* (May 6, 2013). Though Huq's reporting may seem dated, the trend in popularity has continued and the reason, as explained by Sharon Ellerby, writing for the International Association of Professional Writers and Editors, is an increasing appeal to wider youth audiences *and* adult audiences, especially adults with limited time for reading. Ellerby, "Graphic Novels Are Gaining In Popularity," *Writing and Editing Resources* (IAPWE, November, 2017), https://iapwe.org/graphic-novels-are-gaining-in-popularity/.

17. One of the best uses of the comic medium today to develop stinging satirical social commentary is Mark Russell's and Steve Pugh's reimagining of *The Flintstones* for DC Comics.

18. My choices for the graphic novels I focus on were made precisely because of their overt reliance upon apocalyptic narratives, but there are a variety of graphic novels that could easily offer entry points for this analysis while avoiding the apocalyptic narratives. For instance, Paul Chadwick's *Concrete* stories offer ample opportunities to engage with ideas about the environment, environmental responsibilities, and environmental philosophy.

Chapter 1

1. Slavoj Žižek, *Looking Awry* (Cambridge: MIT Press, 1992): 34.

2. Tom Athanasiou, *Divided Planet: The Ecology of Rich and Poor* (Athens: University of Georgia Press, 1998): 75.

3. Claire Curtis, *Postapocalyptic Fiction and the Social Contract* (New York: Lexington Books, 2012). 5. Emphasis mine.

4. Susanna Hoffman, "Anthropology and the Angry Earth: An Overview," 2.

5. Susanna Hoffman, "Anthropology and the Angry Earth: An Overview," 2.

6. This statement is often incorrectly attributed to Slavoj Žižek though Žižek's claim is actually a para-rephrasing of a Fredric Jameson quotation. Beyond Jameson, however, the bon mot most likely finds its origin in H. Bruce Franklin's essay "What Are We to Make of J. G. Ballard's Apocalypse?" in *Voices for the Future: Essays on Major Science Fiction Writers*, vol. 2, ed. Thomas D. Clareson (Bowling Green, OH: 1979): 82–105.

7. Susanna Hoffman, "Anthropology and the Angry Earth: An Overview," 11.

8. For a more in-depth look at the idea of eco-cosmopolitanism see *The Bioregional Imagination: Literature, Ecology, and Place*, eds. Tom Lynch, Cheryll Glotfelty, and Karla

Armbruster (Athens: University of Georgia Press, 2012).
  9. Susanna Hoffman, "Anthropology and the Angry Earth: An Overview," 5.
  10. I address this issue much more fully in *Ecological Reflections on Post-Capitalist Society* (Cornerstone Press, 2018).
  11. Kevin de Laplante, "Making the Abstract Concrete," in *Comics as Philosophy*, ed. Jeff McLaughlin (Jackson: University of Mississippi Press, 2005): 159.
  12. Susanna Hoffman, "Anthropology and the Angry Earth: An Overview," 6.
  13. Peter Frase, *Four Futures: Life After Capitalism* (New York: Verso, 2016): 24.
  14. Greg Rucka and Michael Lark, *Lazarus* (Image Comics, 2013). *Lazarus* is an on-going story collected in five trade paperback volumes to date.
  15. Rick Remender and Sean Murphy, *Tokyo Ghost* (Image Comics, 2016). *Tokyo Ghost* ran through 10 issues and was collected in two paperback volumes.
  16. Anthony Loewenstein, *Disaster Capitalism: Making a Killing Out of Catastrophe* (London: Verso, 2017): 7.
  17. Susanna Hoffman, "Anthropology and the Angry Earth: An Overview," 2.
  18. Daniel Glendening, "Rucka and Lark Reunite for Dystopian Lazarus," *Comic Book Resources* (August 22, 2012), accessed online.
  19. Remender, *Tokyo Ghost*, vol. 1, chapter 2.
  20. Remender, *Tokyo Ghost*, vol. 1, chapter 2.
  21. Rick Remender and Greg Tocchini, *Low* (Image Comics, 2015). This is an on-going series currently collected in 4 volumes.
  22. Hugh Howey, *Wool* (Jet City Comics, 2014). The graphic novel adaptation of Howey's novel leaves a lot to be desired when compared to the novel version. However, the graphic novel raises the same questions as the novel and, for my purposes, is sufficient for extending my analysis on distant future dystopias with apocalyptic aspirations. So, though I think this graphic novel alone, of all the ones I rely on throughout this analysis, is subpar, I think Howey's determination of the apocalyptic possibilities is worthwhile and, therefore, worthy of serious consideration.
  23. There are several well-known underwater facilities, historically speaking, that provide inspiration for today's and tomorrow's attempts at undersea living. For instance, in the 1960s Jacques Cousteau's Conshelf project and a competing project by the United States government, SEALAB, were meant to facilitate underwater research. Those projects, along with a few less successful and lesser known attempts, inspire today's designers. Phil Pauley's Sub-Biosphere, still in development, is meant to be a submersible bio-dome similar to what Remender has envisioned for his underwater inhabitants. There are other designs and aspirations for underwater life, so the technology certainly exists to facilitate Remender's migration of humans to the watery depths.
  24. World Population Prospects, the 2015 Revision, Department of Economic and Social Affairs, United Nations, 2015.
  25. Vivian Cumming, "How Many People Can Our Planet Really Support?" *BBC Earth* (March 14, 2016), accessed online.
  26. L. Bruce Jones of U.S. Submarines is currently developing the undersea Poiseidon Resort and their spin-off H2OMEs that are meant to capture the undersea real estate market.
  27. The best example of this in fiction is Neal Stephenson's *Snow Crash* (New York: Del Rey, 2000).
  28. Mark Jendrysik, "Fundamental Oppositions: Utopia and the Individual," *The Individual and Utopia: A Multidisciplinary Study of Humanity and Perfection*, eds. Clint Jones and Cameron Ellis (Surrey: Ashgate, 2015): 39–40.
  29. Risk Assessment for Toxic Air Pollutants: A Citizen's Guide, Environmental Protection Agency, February 2016.
  30. See the Donora Smog Museum for archives and articles relating to the event as well as an historical overview on their home page.
  31. Damien Carrington, "Pollution responsible for quarter of deaths of young children, says WHO," *The Guardian* (March 7, 2017), accessed online. World Health Organization, "The Cost of a Polluted Environment: 1.7 Million Child Deaths a Year, says WHO" (March 6, 2017), published online, http://www.who.int/en/news-room/detail/06-03-2017-the-cost-of-a-polluted-environment-1-7-million-child-deaths-a-year-says-who. A companion study done by the WHO, "The Top 5 Causes of Death in Children under 5 Years Linked to the Envi-

ronment," breaks that number down into the leading causes and the corresponding number or children affected which reveals a staggering 931,000 children die from causes related to air and water pollution.

32. Another graphic novel that deals with toxification is Johnny Lau's forthcoming *World Water Wars*, however, unfortunately, it had not gone to print in time for me to utilize his work. Still, I think the analysis that derives from the toxicity questions inherent in *Wool* would be applicable to Lau's work though obviously it would entail drawing upon different evidence I think the conclusions would be similar enough that I am convinced my arguments are not weakened by the lack of a graphic work that deals with the toxification of water or soil.

33. Susanna Hoffman, "The Worst of Times, The Best of Times: Toward a Model of Cultural Response to Disaster," "Angry Earth," 134.

34. Bill McKibben offers a wonderful analysis of how problems that seem manageable now could become long-term problems if left unaddressed in his book *Eaarth* (New York: Times Books, 2010). My own analysis follows his closely where the problems of global climate shift are concerned.

35. The best sustained analysis of what the world would look like with no, or very few, remaining humans is Alan Weisman's acclaimed *The World Without Us* (New York: Picador, 2008). I have applied his ideas about a depopulated world to the environments presented in the graphic novels used for analysis here, but I have deviated from or extended his ideas where applicable to make my application of the concept of a people-less world coherent in the context of the differing graphic novels.

36. The distinction I am drawing here is between a secular engagement with apocalypse, denoted hereafter with a lower case "a," and a spiritual engagement with Apocalypse, hereafter denoted with an uppercase "A." It is important to separate the two because each entails a particular set of beliefs and responses to the apocalyptic future scenario and those responses not only fail to overlap in important ways, but they can be antagonistic as well. Given that my focus is primarily on the secular conception of the apocalypse I want to carefully separate the two imaginative approaches to the future.

37. Genesis, 2:15. I am using the English Standard Version for all Biblical material.

38. The meaning of dominion in this context is hotly contested, but it is a spirited debate which need not concern us here. However, J. Baird Callicott's *Earth's Insights* provides a thorough scholarly explanation of the debate and, developing Callicott's analysis, Marshall Jolly and Clint Jones, "A New Christian Context for Appalachia," *Journal of Appalachian Studies*, provides a critical evaluation of the debate in terms of applicability to environmental action.

39. Genesis, 2:16.

40. Genesis, 2:25 says, "the man and his wife were both naked and were not ashamed." Genesis, 3:7 says, "Then their eyes were opened, and they knew that they were naked."

41. Genesis, 3:24.

42. Genesis, 3:21.

43. Genesis, 6:5–7.

44. Jason Aaron, *The Goddamned* (Image Comics, 2017).

45. *The Goddamned*, vol. 1, ch. 4. Adam and Eve's son, Cain, is the narrator and main character of Aaron's story.

46. Genesis, 6:9.

47. Genesis, 6:13.

48. Genesis, 6:11–12.

49. This calculation is based upon the extrapolations provided in "Chronology of Noah's Time in the Ark," 63, of the ESV Bible I am using—see bibliography for full citation.

50. Plato mentions "great floods" or "deluges" in his *Laws*, *Timaeus*, and *Critias*. Greek culture uses the device of a great flood three times, the flood of Ogyges, the flood of Deucalion, and the flood of Dardanus. The former two are thought to have signaled the end of particular ages in human history, the Silver and First Bronze Ages.

51. Maggie Nelson, *The Art of Cruelty* (New York: W.W. Norton, 2011): 134.

52. For an excellent interrogation of how effective the plague was in killing off human populations see Robert Gottfried, *The Black Death: Natural and Human Disaster in Medieval Europe* (London: Robert Hale, 1983), especially chapter 3. Gottfried contends that though most scholars put the mortality rate around 1 in 3 persons this is somewhat inaccurate to say nothing of misleading. Certain places in Italy, France, and Spain suffered mortalities in the 50–60 percent range and

some sub-populations suffered 70 percent or greater mortality. The Black Death is as close to an actual apocalyptic extinction event as human society as ever gotten on a global scale.

53. George Deaux, *The Black Death, 1347* (London: Hamilton, 1969): 145.

54. A. Lloyd Moote and Dorothy Moote, *The Great Plague: The Story of London's Most Deadly Year* (Baltimore: Johns Hopkins University Press, 2004): 263.

55. Cholera has claimed the lives of millions of people to date, and outbreaks are regular occurrences in non-industrial countries. Smallpox, though subject to a successful vaccine, and having been declared eradicated in 1980, shortly after the last known death to occur from smallpox in 1977, is representative of the fears people have of poxes in general. According to the WHO several poxes remain a viable threat though they have been less successful as pandemics so far; these include monkeypox, ORF virus, and molluscum contagiisum. Flu remains the seventh largest killer in the United States and in 1918 the Spanish Flu infected more than 500 million people and killed an estimated 50 million people. The incurable nature of HIV/AIDS keeps it at the fore of public fears. E. coli is a constant threat having produced 20 outbreaks since 2010 according to the CDC though the outbreaks have not managed to take on the global level threat that other epidemics in the past have achieved. See Moote and Moote's epilogue in *The Great Plague* for a thorough treatment of future plague possibilities not limited to the bubonic variety.

56. Ursula Le Guin develops the idea of the yin/yang nature of utopia and dystopia in a collection of essays that accompany the 2016 edition of Thomas More's *Utopia* issued by Verso Press. The notion of a yin/yang understanding of utopia is developed over the course of the four essays, but particularly the second, "Utopiyin, Utopiyang."

57. Homi Bhabha, *The Location of Culture* (London: Routledge, 1994; Routledge Classics, 2004).

58. Ali Brox, "'Every Age Has the Vampire It Needs': Octavia Butler's Vampiric Vision in Fledgling," *Utopian Studies* 19, no. 3 (2008): 391.

59. Lauren Berlant, *Cruel Optimism* (Durham: Duke University Press, 2011): 10.

Chapter 2

1. Susanna Hoffman, "Anthropology and the Angry Earth: An Overview," 7.

2. Puspa Damai, "Messianic-City: Ruins, Refuge, and Hospitality in Derrida," *Discourse* 27, no. 2/3 (Spring and Fall 2005): 70.

3. Puspa Damai, "Messianic-City: Ruins, Refuge, and Hospitality in Derrida," *Discourse* 27, no. 2/3 (Spring and Fall 2005): 68.

4. To be fair this trichotomous formulation could be rendered differently by stating that the voter apathetically votes for Trump because they hold conservative values—regardless of whether those values are in line with Trump's politics—and the future of a Clinton presidency is a further degradation of society into a welfare state and Sanders is the undoing of society. It does not take much tinkering to see how the situation of the voter disaffected by a long history of frustration caused by the seemingly unending failures of utopian yearnings and projects.

5. Alec Charles, "The Meta-Utopian Metatext: The Deconstructive Dreams of Ulysses and Finnegan's Wake," *Utopian Studies* 23, no. 2 (2012): 477.

6. Paul Riceour, *Lecturers on Ideologies and Utopia* (New York: Cambridge University Press, 1986): 1–2.

7. Appalachian Voices, "Mountain Top Removal 101," www.appvoices.org, updated 2013, last accessed February 22, 2015.

8. David Kohn, "The 300 Million Gallon Warning," *Mother Jones*, March/April 2002. www.MotherJones.com, updated 2015.

9. David Kohn, "The 300 Million Gallon Warning."

10. U.S. Energy Information Administration, "Petroleum and Other Liquids," www.eia.gov, last updated June 2014.

11. Global Climate Shift is commonly mislabeled as Global Warming which has had the unfortunate side effect of causing widespread misunderstandings regarding what is happening to our climate.

12. These estimates are derivative of several arguments but align primarily with David Pimentel and Marcia H. Pimentel, *Food, Energy, and Society*, 3rd edition (Boca Raton: CRC Press, 2008). This issue of foraging for survival is much more complex than I have outlined here because of the various layers of social development and interaction

that groups would create or encounter. For more on how to model these issue to develop more concise interpretations of the required area necessary to survival see Jacob Freeman and John Anderies, "The Socioecology of Hunter-gatherer Territory Size," *Journal of Anthropological Archeology* 39 (2015): 110–123.

13. One World Nations Online, "The Most Populated Cities in the United States," www.nationsonline.org, last updated December 2002.

14. Lei Chen, Randall Todd, Julia Keihlbauch, Maroya Walters, Alexander Kallen, "*Notes from the Field*: Pan-Resistant New Delhi Metallo-Beta-Lactamase-Producing *Klebsiella pneumoniae*—Washoe County, Nevada, 2016," *Morbidity and Mortality Weekly Report* 66, no. 1 (2017): 33, accessed online.

15. Susan Brink, "A Superbug That Resisted 26 Antibiotics," *NPR* (January 17, 2017), accessed online.

16. Susan Brink, "A Superbug That Resisted 26 Antibiotics," *NPR* (January 17, 2017), accessed online. However, a Reuters investigation into the CDC estimate alleges that the CDC relied too heavily on small sample sizes, old data, and too few geographic areas. Ryan McNeill, "Deconstructing the CDC's 'snapshot' estimates," n.d., accessed online.

17. Very rapid could be instantaneous, but it would be unlikely except in instances of nuclear war, perhaps, but a reasonable rapid decimation of the population could be expected to take months. This alone represents a problem because worsening conditions would likely send large numbers of people to population centers as they seek government assistance and available resources. However, their migrations would leave developed rural areas unattended long before a critical mass of 99.99 percent was reached. Hence, many of these problems facing the post-apocalyptic survivor could potentially be underway long before people became aware of the necessity to acknowledge them.

18. The vagueness of this statement hinges on the fact that there are lots of radioactive substances and they all decay at different rates ranging across a half-life spectrum from 30 to 24,000 years. For more, see United States Nuclear Regulatory Commission, "Report: Backgrounder on Radioactive Waste," https://www.nrc.gov/reading-rm/doc-collections/fact-sheets/radwaste.html.

19. U.S. Energy Information Administration, "How Many Nuclear Power Plants are in the United States and Where Are They Located?" www.eia.gov, last updated January 22, 2015.

20. "Devil in the Deep Blue Sea," *on Earth*, November 11, 2014, https://www.nrdc.org/onearth/devil-deep-blue-sea. A hypoxic zone is an area of low oxygen concentration that is capable of suffocating and killing animals and some plant life which is why they are referred to as "dead zones" and Diaz estimates that a more accurate count is 1,000-plus dead zones globally.

21. Susanna Hoffman, "Anthropology and the Angry Earth: An Overview," 3.

Chapter 3

1. Robert Kirkman, *The Walking Dead Compendiums 1, 2, 3* (Image Comics, 2009, 2012, 2015, respectively). For citation purposes, since the compendiums do not have page numbers, I will cite passages and frames of the comic using notation that indicates which compendium and chapter. So, for instance, discussing the beginning of the apocalypse, I would cite compendium 1, chapter 1 as Kirkman, 1.1.

2. At this point, and generally throughout my analysis, I am intentionally ignoring the plot lines of the popular television version of *The Walking Dead* as well as its prequel spinoff *Fear the Walking Dead* though I may allude to them minimally if necessary, however, *The Walking Dead* is a massively successful franchise with storylines developed in console and smart phone video games and literature in addition to its cinematic counterparts. My focus is squarely on the comic and collected graphic novel versions and I realize that some of the aspects of the story I am dealing with have been "explained" or otherwise fleshed out in one of these other formats. Still, I believe that the original story is not only the best, but the best for my purposes here, so the necessity to deviate from that analysis to address one of these other formats would have to be great indeed.

3. Kirkman, 1.1.
4. Kirkman, 1.1.
5. Kirkman, 1.1.

6. Kirkman, 1.1.
7. Kirkman, 1.1.
8. Kirkman, 1.1.
9. Kirkman, 1.1.
10. Kirkman, 1.1.
11. Kirkman, 1.2.
12. Kirkman, 1.2.
13. "http://www.iweathernet.com/educational/warm-snowstorms-and-atlanta-biggest-snowfalls-on-record," regularly updated, last accessed December 19, 2016. iWeatherNet.com is a NOAA WRN Ambassador.
14. Kirkman, 1.2.
15. Kirkman, 1.2.
16. Kirkman, 1.2.
17. Kirkman, 1.2.
18. Kirkman, 1.2.
19. Kirkman, 1.3. Emphasis added.
20. *The Walking Dead* provides a lot of material to work with in terms of how civilization might reform, or what aspects of society might be the focus of a large group of survivors, and I have addressed some of these in a previous book, especially the role of marriage in small groups and large societies. For more, see *A Genealogy of Social Violence* (Surrey: Ashgate, 2013).
21. All of the scenes pertaining to this particular journey can be found in Kirkman, 1.3.
22. Kirkman, 1.7.
23. Kirkman, 1.7.
24. Kirkman, 2.9.
25. Kirkman, 2.9.
26. Kirkman, 2.10.
27. www.southeastcoalash.com provides a comprehensive, interactive map of the locations of such sites as slurry ponds, sludge ponds, and landfills. Tracing out different routes for the group to take would not limit their coming into contact with these sites since there are hundreds of them just in the southeast where the action of *The Walking Dead* takes place.
28. http://earthjustice.org/features/coal-ash-contaminated-sites# also features an interactive map which displays the locations of coal related sites.
29. https://www.npms.phmsa.dot.gov/default.aspx provides an interactive map showing the areas, by county, where pipelines exist. http://www.pipeline101.com/where-are-pipelines-located provides a full scale map of the United States showcasing the locations of all major pipelines moving crude and refined oil as well as natural gas throughout the United States.
30. https://www.nrc.gov/info-finder/reactors/ for more information on nuclear facilities in the areas mentioned.
31. http://www.motherjones.com/environment/2014/04/west-texas-hazardous-chemical-map provides an interactive map showing the locations of all such plants in the United States.
32. Jaeah Lee, "Map: Is There a Risky Chemical Plant Near You?" http://www.motherjones.com/environment/2014/04/west-texas-hazardous-chemical-map.
33. http://www.energyjustice.net/map/localmap provides a high quality, interactive map that tracks not just the most recognizable threats—the ones I have been focusing on up to this point—but a comprehensive list that includes: nuclear, coal, oil, gas, oil refinery, LNG (natural gas), ethanol, trash, biomass, poultry, tire, landfill, landfill gas, hydro, geothermal, wind, solar, sewage incineration, and others across a spectrum that identifies operating, proposed, expanded, and closed facilities.
34. http://www.energyjustice.net/map/localmap these arguments omit the 53 sites within 50 miles of Richmond, the 150 sites within 50 miles of Fayetteville, NC, or the 59 sites in the Columbia/Florence, SC, area, all of these areas are regions that the group would have traveled through from Atlanta to follow the I-95 corridor north, and following that corridor to D.C. This also ignores the 60 or so sites surrounding Cynthiana and Lexington, the 43 sites that exist within a 50 miles radius of Knoxville, or the 43 around Chattanooga, which represent roughly 150 hazardous industrial locations along the I-75 corridor, a route Rick traveled three times within a year and half of the apocalypse starting.
35. Kirkman, 2.12.
36. Kirkman, 2.12.
37. Kirkman, 2.14.
38. Kirkman, 2.14.
39. Kirkman, 2.15.
40. Kirkman, 2.16.
41. Kirkman, 3.19.
42. Kirkman, 2.16.
43. Kirkman, 3.18.
44. Kirkman, 3.19.
45. Kirkman, 3.21.
46. Kirkman, 3.23.
47. Kirkman, 3.24.

48. Kirkman, 3.19.

49. This information is compiled from various sites and the numbers reflect aquariums as well as zoos since both are classified similarly and information for them is consolidated. For more information see http://www.statisticbrain.com/zoo-statistics/, Laura Fravel "Critics Question Zoos Commitment to Conservation," http://news.nationalgeographic.com/news/2003/11/1113_031113_zoorole.html, and http://www.americanzoos.info/.

50. Peter Paik, *From Utopia to Apocalypse* (Minneapolis: University of Minnesota Press, 2010): 82.

51. Kirkman, 3.24.

52. For a superb analysis of the role of the frontier mythos in popular culture, including an insightful understanding of the role of nature, see John Moeller, "Nature, Human Nature, and Society in the American Western," *Political Mythology and Popular Fiction*, eds. Ernest Yanarella and Lee Sigelman (New York: Greenwood Press, 1988). Though Moeller focuses his attention on the western genre his analysis parallels my arguments here nicely.

53. Peter Sloterdijk, *Bubbles* (Los Angeles: Semiotext(e), 2011): 164.

Chapter 4

1. Brian K. Vaughn and Pia Guerra, *Y: The Last Man*, deluxe editions 1–5 (Vertigo, 2008–2011). Unlike the compendium format of *The Walking Dead*, which lacks proper pagination, *YTLM* does use page numbers throughout so I will be citing references to these graphic novels as edition and page, hence, edition 1 page 90 would be (1.90).

2. An attempt to explain the origins of the zombie virus is made in the second season of the television spin off when the group detours to the CDC. One lone scientist on the verge of suicide is able to give some insight into the virus and the way it works, though his explanations are still somewhat cryptic leaving the real causes and potential of the virus up to the imagination of the viewer. A similar attempt is made in the subsequent prequel to the flagship Walking Dead series. However, no such attempt is made in the graphic novel.

3. Vaughn, 1.12.
4. Vaughn, 1.35.
5. Vaughn, 1.36.
6. Vaughn, 1.37.
7. Vaughn, 1.39.

8. Flightradar24.com allows you to view the flights in air in real time from individual airports to the entire globe. The sheer number of flights off the ground above the U.S. is mind-boggling, to be honest, and this website can only track flights that are equipped with a special kind of transponder that not all planes utilize. This makes the numbers of flights in the air larger than actual flight data would suggest, especially outside the United States where planes are even less likely to utilize the transponder. Flightradar24 is still, in spite of that, considered the best source for real time flight data.

9. Laura Bloom, "Why Aren't There More Female Airline Pilots? This High-Flying Woman Is Breaking Boundaries," www.forbes.com, 23 June 2016.

10. Department of Aviation, O'Hare International Airport, "Monthly Operations, Passengers, Cargo Summary By Class," published as public record on the O'Hare website. I chose the month of August 2016 for two reasons: first, because the storyline suggests that late summer early fall is when the cataclysmic event happens, and second, I wanted to avoid a holiday month with higher rates of travel. I also wanted not only the most up-to-date information, but also the most complete data set available, so I used 2016 data.

11. Vaughn, 1.144.
12. Vaughn, 3.23.
13. Vaughn, 3.24.
14. Vaughn, 5.298.

15. The number of total ships in the global fleet is compiled by Statista.com but to get a real time look at the number of ships moving about the globe at any given time the website MaritimeTraffic.com provides the same real time experience as flightradar24.com and it is easy to see that the number of ships on the oceans and seas is in excess of that 51,000 ship total.

16. Heather Long, "Female CEO's are at a record level in 2016, but it's still only 5%," www.CNN.Money.com, September 29, 2016. This is up from Vaughn's estimate, but only slightly.

17. Total Labor Force, data.worldbank.org, last updated 2016 but confirmed totals and complete information available is from 2014.

18. Lynn Langton, "Crime Date Brief: Women in Law Enforcement 1978-2008" *Bureau of Justice Statistics* (June 2010). The data is compiled using the most current census data available.
19. Vaughn, 1.41.
20. "Are there disease risks from dead bodies and what should be done for safe disposal?" The World Health Organization, 2017.
21. Vaughn, 1.53.
22. Vaughn, 1.65.
23. Vaughn, 1.106.
24. Vaughn, 1.97.
25. Vaughn, 1.114.
26. Vaughn, 1.115.
27. Vaughn, 1.131.
28. Boston, like Chicago, is home to a very large airport, Logan International, which is the largest airport in New England and the seventeenth busiest airport in the United States. Yet, despite being a large international airport, there is no evidence that a single plane crash landed in Boston. Boston proper is home to 646,000 people while the greater Boston metropolitan area houses 4.7 million people. Boston is the tenth largest city in the United States. Curiously, though Waverly in D.C.—a city with only about 4,000 more people than Boston—is still picking up corpses, none of the depictions of Boston have dead bodies in them, evidence of dead bodies, or people picking them up.
29. Vaughn, 1.133–146. This sequence of events includes a narrative detour that focuses on the Amazons in Boston hunting for Yorick. The scene is short, but it is located outside Fenway Park and, as with the rest of Boston and the surrounding area, the city looks an awful lot like a pre-Apocalypse Boston.
30. Vaughn, 1.134 shows a smattering of people attempting to board various trains. But in his conversation with the rail worker Yorick bribes to secure passage on the train there is no real hint that, aside from the lack of luxuries, life in Boston is terrible.
31. Vaughn, 2.220.
32. Vaughn, 1.145.
33. Vaughn, 1.184.
34. Vaughn, 1.191.
35. Vaughn, 1.206.
36. Vaughn, 1.213.
37. Vaughn, 2.10.
38. Vaughn, 2.35.

39. World Statistics, "Nuclear Energy Around the World," www.nei.org, November 2016. To see a map of all nuclear facilities worldwide, including those active, offline, shutdown, or being built, visit www.carbonbrief.org.
40. Vaughn, 2.58.
41. Vaughn, 2.129.
42. Vaughn, 2.172.
43. Vaughn, 2.180.
44. Vaughn, 2.242.
45. Vaughn, 2.246.
46. Vaughn, 2.245.
47. Vaughn, 3.80
48. Vaughn, 3.187.
49. Laura Parker, "Ocean Trash: 5.25 Trillion Pieces and Counting, the Big Questions Remain," *National Geographic* (January 11, 2015), news.nationalgeographic.com. Parker explains the origins of these numbers in the article saying, "The process of collecting and counting is meticulous, time-consuming work. It took Marcus Eriksen, co-founder of the 5 Gyres Institute, a nonprofit ocean advocacy group, more than four years, using samples gathered from 24 survey trips, to come up with his estimate that 5.25 trillion pieces of debris float on the surface."
50. Smithsonian, *National Museum of Natural History*, "Ocean Trash Plaguing Our Sea," ocean.si.edu, last updated May 31, 2015.
51. Jessica Hartogs, "Three Years After Oil Spill, Active Clean-up Ends in Three States," *CBSNews* , June 10, 2013, www.cbs.com.
52. "Number of Tankers in the World's Merchant Fleet in 2010, by Vessel Type," www.statista.com. In 2010, the number was 3,948.
53. "Transporting Oil by Sea," www.planete-energies.com, July 7, 2015.
54. "Transporting Oil by Sea," www.planete-energies.com, July 7, 2015. LNG is liquefied natural gas and is predominately methane which makes the gas easier to transport but, in addition to being odorless, colorless, and non-corrosive, is toxic.
55. Vaughn, 3.189.
56. Vaughn, 3.191.
57. Vaughn, 3.212. Most of the bird-eye views of Captain Kilina's ship are small frames, but the submarine, page number referenced here, offers a view of the ocean that is clean, with a healthy school of fish going about their business. There is nothing

in this, or the other portrayals of the ocean, during the voyage, in particular, that would suggest an apocalypse.

58. Vaughn, 3.274.

59. Though the focus on Hero's journey is substantially more limited than the time given to Yorick's travels through the various regions when Hero finally returns to Kansas the corn fields are planted, and food production on a massive scale would seem to be underway. Though the authors here make no effort to confirm or deny the state of the union as it were, it is clear that the more organized elements of society are starting to create stability and a sense of normalcy. Vaughn, 4.168 and after depict the cornfields and other aspects of Hero's journey that provide the evidence for this assertion.

60. Vaughn, 4.135 and 4.145 contain the relevant frames.

61. After leaving Australia the group picks up a fourth person, one of the Australian submariners that is infiltrating the group to keeps tabs on the last man, but who also has romantic ties to Dr. Mann. When the group arrives in Japan they split up in search of Yorick's monkey and Dr. Mann's mother. Dr. Mann and Rose, the Australian sailor, head out on foot to seek out Dr. Mann's mother in Yokogata (4.160) while Yorick and Agent 355 head to Tokyo via the rail system, which provides a stunning, and stunningly clean, view of the countryside (4.152).

62. Vaughn, 4.192.

63. Consider Yorick's view from his train car as he and Agent 355 leave Beijing, Vaughn 5.141.

64. Women currently only make up 10 percent of the railroads workforce in the United States and the majority of those women are in support and administration. Deborah Huso, "Women on the Line: Railroads still lag when it comes to hiring women, but Class I's Aim to Pick Up the Pace," *Progressive Railroading*, www.progressiverailroading.com, August 2015. Though the industry is making efforts to hire more women and diversify the industry the likelihood that there would be a lot of women on trains at the time of the gendercide is empirically low. Many rail systems outside of city transit and airport shuttling are not fully autonomous requiring a human element to operate the train especially in instances of obstacles on the tracks. If the proposed gendercide occurred there is a high probability that the railroad industry would suffer a magnitude of loss and damage equitable to the airline industry. As such, going all the way back to the beginning of the story when Yorick first leaves Boston and, later, travels from Marysville to Missouri, it is unlikely that the rail systems would be able to foster that amount of travel; this is to say nothing of the decrease shipment of goods across the country. Later, when Yorick is traveling in Japan and using the Tran-Siberian railroad to get from China to France, it is possible that these trains are able to function because, one, Asian and European railway systems are more automated and, two, by the time Yorick gets to these places several years have passed which would allow for these societies to clean-up any wreckage and get the railroads operational again.

65. Vaughn, 5.177. Though only briefly addressed, the Israeli troops pursuing Yorick steal the mail in Paris, sacks of mail, to try to gain intel on his whereabouts.

66. Vaughn, 5.264.

Chapter 5

1. Jeff Lemire, *Sweet Tooth* (Vertigo, November 2009–February 2013) was originally published as a monthly comic that ran for 40 issues. It has since been collected into a three-volume library edition set, but I will be using the comics so I will make references to the issues themselves since they do not have pagination. Hence (Lemire, 1) would be a reference to the first issue, not a page.

2. Paul K. Saint-Amour, *Tense Futures* (New York: Oxford University Press, 2015): 136.

3. Lemire, 35.

4. Lemire, 2.

5. Lemire, 3.

6. Though there are several references to the virus in the various flashback sequences throughout the series only one person is shown to suffer the disease from start to finish as a part of the present-tense storyline. Though the disease is painful and progressively debilitating, it does not prevent the individual from continuing to function not only with a high level of mental acuity but with a high degree of physical activity for a substantial amount of time. If the depic-

tion of this person's suffering is typical of what most people suffered though, then it is entirely possible that most of the urgent aspects of social stability could have been addressed even if those things had to be done during riots and general panic.

7. Lemire, 40.

8. *Snowpiercer* is the umbrella label for the overarching story, and while it is also the name of the train in the first graphic novel, it is a mistake to associate the world the story occurs in with the train that bears the same name. That is, the world of *Snowpiercer* is different than the train Snowpiercer and I will use the convention of italicizing the world but not the train for the sake of clarity. Further, in order to maintain citation clarity given the disjointed nature of the story, and its compilation, I will use the subtitles for references to particular story details and will utilize the authors name to indicate the version of the story. The original story was written by Jacques Lob and subtitled "The Escape," the second installment was written by Benjamin Legrand and subtitled "The Explorers," the third, and most recent installment, was written by Olivier Bocquet and was given "Terminus" as its subtitle. I say the third installment is the most recent because Legrand's contribution was done in two parts originally labeled 2 and 3, which was subtitled "The Crossing" but was included with Legrand's "explorers" as a single volume. However, in the second installment the integrity of the story is maintained as Legrand's contributions are kept separate. For my purposes, when citing Lob, Legrand, or Bocquet I will use a traditional name and page number (for example, Lob, 1). However, to make general references to the different versions I will use their subtitles.

9. Jacques Lob and Jean-Marc Rochette, *Snowpiercer* (Titan Comics, [1984] 2014). Originally published in French as *Le Transperceneige* in 1982, all three versions of the *Snowpiercer* saga that I am referring to were published by Titan Comics and issued as hardbound library copies beginning in 2014. Legrand's compiled volume was published in 2014 while Bocquet's was published in 2016. It is important to note that the internal inconsistency of the story notwithstanding all three contributions were not produced in pure artistic isolation. Though Jacques Lob passed away in 1990, artist for the series Jean-Marc Rochette has worked on all three volumes.

10. Carl Saga and Paul Erlich, *The Cold and the Dark: The World After Nuclear War* (New York: W.W. Norton, 1985). This is also the same year that Jacques Derrida's "No Apocalypse, Not Now (full speed ahead, seven missiles, seven missives)," trans. Catherine Porter and Philip Lewis, appeared in *Diacritics* 14.2 (Summer 1984). Subsequently there has been a considerable amount of major statements and analysis about the nuclear age including, recently, *Diacritics* 41.3 (Fall 2013) which is an issue marking the thirtieth anniversary of the original colloquium that produced Derrida's essay.

11. Lob, 3.
12. Lob, 5.
13. Lob, 5.
14. Lob, 15.
15. Paul K. Saint-Amour, *Tense Future* (New York: Oxford University Press, 2015): 25.
16. Lob, 16.
17. It is quite likely that this depiction is the inspiration for the animated prequel short film (2013) which was released ahead of the full length movie version.
18. Lob, 18.
19. A firestorm is a conflagration that burns so intensely that it produces its own self-sustaining storm level winds and moves itself away from the area of its origin by consuming fuels outside the scope of the initial flames.
20. Lob, 64.
21. Lob, 65.
22. Lob, 65.
23. Lob, 65.
24. Robert Rodden, Floyd John, and Richard Laurino, "Exploratory Analysis of Firestorms," *Office of Civil Defense, Department of the Army* (May 1965): 35.
25. Becky Oskin, "What Would Happen if Yellowstone's Supervoclano Erupted?" *Life's Little Mysteries*, www.livescience.com, May 2, 2016.
26. Becky Oskin, "What Would Happen if Yellowstone's Supervolcano Erupted?"
27. Lob, 41.
28. Legrand, 10.
29. Legrand, 16.
30. Legrand, 45.
31. Legrand, 67.
32. Legrand, 69.
33. Legrand, 63.
34. Legrand, 120.

35. Scott McCloud, *Understanding Comics*, 63.
36. Legrand, 135.
37. Legrand, 134.
38. Bocquet, 4.
39. Bocquet, 5.
40. Bocquet, 6.
41. Bocquet, 7.
42. Bocquet, 9.
43. Bocquet, 26.
44. Bocquet, 29.
45. Bocquet, 70.
46. Bocquet, 86.
47. Bocquet, 91.
48. Bocquet, 92.
49. Bocquet, 100.
50. Legrand, 130.
51. Bocquet, 105-7.
52. Bocquet, 165.
53. Bocquet, 166.
54. Bocquet, 208.
55. Bocquet, 210.
56. Bocquet, 217.
57. Bocquet, 220.
58. Bocquet, 222.

Chapter 6

1. Wood, 1.5.
2. Brian Wood, *The Massive*, vols. 1–5, Dark Horse Comics, 2013–2015. Though *The Massive* was originally published as a comic book series it has been collected in its entirety into five volumes which I will use instead of the comics. Each volume, unlike other graphic novels, contains proper pagination, and, as such, will be cited throughout using volume and page (i.e., Wood, 1.1, for volume one, page one). In addition to the five volumes that compose the original story, Brian Wood and the creative team responsible for *The Massive*, also published a prequel in 2016 which I will cite as volume zero (Wood, 0.1).
3. Susanna Hoffman, "Anthropology and the Angry Earth: An Overview," 4
4. As I did with *Sweet Tooth* I am ignoring the extraterrestrial and supernatural subtext of Wood's story especially because that theme is unrevealed until the end of the story. Wood clearly intends to display a possible future fully caught in the destructive maw of runaway global climate shift.
5. Christian Parenti, *Tropic of Chaos* (New York: Nation Books, 2011): 7.
6. Wood, 0.7. Emphasis added.
7. Wood, 0.110.
8. Wood, 3.5.
9. Wood, 1.13. This is the date ascribed to the Cook Island storms at least twice in the story, however, an alternative timeline, beginning on the winter solstice, is suggested by Callum Israel after he learns key information about Mary and her relationship to the Crash, see Wood, 4.76.
10. Wood, 1.13.
11. Wood, 1.13.
12. Wood, 1.44.
13. Wood, 1.14.
14. Wood, 1.52.
15. Wood, 1.14.
16. Daksha Rangan, "A Massive Chunk of Antarctic Ice is about to Break Off," *Science and Environment*, www.theweathernetwork.com, January 8, 2017.
17. Wood, 1.19.
18. Wood, 1.21.
19. Wood, 1.26.
20. Wood, 1.58.
21. Benjamin Yeh, "Rising Sea Levels Threaten Taiwan," *Taipei Times* (May 10, 2010) accessed online, http://www.taipeitimes.com/News/taiwan/archives/2010.
22. Benjamin Yeh, "Rising Sea Levels Threaten Taiwan." Emphasis added.
23. Nicholas Freschi, "Taiwan's Nuclear Dilemma," *The Diplomat* (March 14, 2018), accessed online, https://thediplomat.com/2018/03/taiwans-nuclear-dilemma/.
24. Wood, 1.38. Though the science behind this makes this occurrence, as Wood describes it, unlikely, it could happen. Methane becomes solid at pressures like those at the bottom of the ocean. The deposits there are also ice so when those ice deposits break away, or in the case of warm seas, thaw, the methane will burble to the surface. Though this has not been witnessed there is evidence to suggest ships have been lost at sea this way. It is unclear how big the methane bubble would have to be, but what is known is that the ships need to be in close proximity to the bubble. Too far away and they may rock bit, but they won't sink, and if they are directly over the bubble its bursting will not sink the ship either. Wood's portrayal of the battle group going down is accurate to the science. For a clear explanation see The Associated Press, "Could Methane Bubbles Sink Ships?" *NBC News* (October 21, 2003), accessed online, http://www.nbcnews.com/id/3226787/ns/

technology_and_science-science/t/could-methane-bubbles-sink-ships/.
  25. Wood, 1.62.
  26. Wood, 1.52.
  27. Wood, 1.92.
  28. Wood, 2.11.
  29. Wood, 2.16. ULCCs are Ultra Large Crude Carriers, otherwise known as Supertankers, and are the largest operating cargo vessels in the world. ULCCs can be in excess of 250,000 Dead Weight Tonnage (DWT) and are capable of carrying huge amounts of crude oil in a single trip from the Persian Gulf to countries in Europe, Asia and North America. FPSOs are Floating Production, Storage, and Offloading ships that can receive hydrocarbons in a variety of ways and then produce oil, store it, and then offload it into a tanker or pipeline system. Many oil operations prefer FPSOs because they are cheaper to build and maintain than platforms. Both vessels can hold huge amounts of oil so postulating that these ships were being torched to barricade the Straits betokens a level of destruction that boggles the mind.
  30. Wood, 1.79, 1.80, 1.86, and 1.98 all show the ocean as Mary swims and one could easily mistake her for a tourist about to get a diving lesson.
  31. Wood, 1.79–1.100 contains the Mogadishu storyline.
  32. Wood, 1.95.
  33. Wood, 2.94.
  34. Wood, 2.112.
  35. Wood, 1.62.
  36. Wood, 4.86.
  37. Christian Parenti notes that just the 22 Pacific Island nations are home to 7 million people (*Tropic of Chaos*, 7) but according to National Oceanic and Atmospheric Administration (NOAA) in 2010 roughly 39 percent of the world's population lived in coastal areas totaling approximately 123.3 million people (https://oceanservice.noaa.gov/facts/population.html) and this number is expected to rise by at least 8 percent by 2020 so it is a fair estimate that the climate crisis combined with conflicts could easily push 200 million people into refugee status.
  38. Wood, 1.34.
  39. Wood, 1.84.
  40. Wood, 4.84.
  41. Wood, 1.87. The price of shark fin soup can actually vary widely between a few dollars and few thousand dollars depending on a lot of factors, but most especially, the type of shark.
  42. Wood, 1.89.
  43. Max Fisher, "This map shows where the world's 30 million slaves live. There are 60,000 in the U.S.," *The Washington Post* (October 17, 2013).
  44. Lemire covers this in issues 3 and 4.
  45. www.freedomunited.org.
  46. A large dam is defined as a dam with a height exceeding 15 meters, that is, taller than a 4-story building, and a major dam is at least 10 times that height, but can also be identified by reservoir volume or dam volume. The flooding a breached or broken dam can cause has the potential to kill tens or hundreds of thousands of people depending on the location of the dam. The weight of the water held behind dams can also cause unexpected seismic activity. https://www.internationalrivers.org/questions-and-answers-about-large-dams/.
  47. The amount of water is provided by the Bureau of Reclamation. https://www.usbr.gov/main/about/fact.html the context for that amount of water is provided by Robyn Brinks Lockwood and Kelly Sippell, *Four Point Reading and Writing Intro*, https://www.press.umich.edu/pdf/ which also includes an analysis of what would happen if the dam broke—an analysis I have closely followed here.
  48. David Orr, "Floodgates of Terror," *Earth Island Journal*, http://www.earthisland.org/journal/index.php/magazine/entry/floodgates_of_terror/ (accessed November 23, 2018).
  49. Eric Fish estimates the amount of water to be the equivalent of 280,000 Olympic size swimming pools. An Olympic size swimming pool contains approximately 2.5 million liters of water, or roughly 660,000 gallons which, when multiplied by Fish's estimation provides the approximate amount I am using here. Eric Fish, "The Forgotten Legacy of the Banqiao Dam Collapse," *The Economic Observer* (February 8, 2013), http://www.eeo.com.cn/ens/2013/0208/240078.shtml.
  50. Wood, 5.142.

Chapter 7

  1. Walter Benjamin, *Illuminations* (New York: Shocken Books, 2007): 242.

2. Timothy Morton, *Ecology Without Nature* (Cambridge: Harvard University Press, 2007): 10.

3. Eben Kirksey, *Emergent Ecologies* (Durham: Duke University Press, 2015): 44.

4. This type of wasteland approach is popular in apocalyptic graphic novels, especially those like *Wasteland* or *WinterWorld* where the landscape is sun bleached desert or a global deep freeze. The pitfall inherent in these, and similar, stories is that they skip the apocalypse and its immediate aftermath and tend to put the story so far into the future that it does not provide a substantial entryway into the type of analysis offered here—these problems are analyzed more fully in chapter 2.

5. Bill McKibben, *Eaarth: Making a Life on a Tough New Planet*, 145. Emphasis added.

6. Christian Parenti, *Tropic of Chaos* (New York: Nation Books, 2011): 224.

7. Timothy Morthon, *Ecology Without Nature*, 11.

8. Timothy Morton, *Ecology Without Nature*, 17.

9. See my *Ecological Reflections on Post-Capitalist Society* (Cornerstone Press, 2018) and *Stranger, Creature, Thing, Other: Moral Monstrosity and Our Ecoestential Crisis* (Cornerstone Press, 2019).

10. Warwick Fox, "Developing a General Ethics (with Particular Reference to the Built, or Human-Constructed, Environment," in *Environmental Ethics: The Big Questions*, ed. David Keller (Oxford: Wiley-Blackwell, 2010): 213–217. The essay in Keller's text is an excerpt of Fox's original essay which can be found in *A Theory of General Ethics: Human Relationships, Nature, and the Built Environment* (Cambridge: MIT Press, 2006).

11. Kirksey, 34.

12. Kevin de Laplante, "Making the Abstract Concrete," in *Comics as Philosophy*, ed. Jeff McLaughlin (Jackson: University Press of Mississippi, 2005): 169–171.

13. Timothy Morton, *Ecology Without Nature*, 185.

14. David Harvey, *Spaces of Hope* (Oakland: University of California Press, 2000).

15. Garret Eckbo, "Urban Nature," *The Town Planning Review* 56, no. 2 (April 1985): 223.

16. Gordon MacLeod and Kevin Ward, *Spaces of Utopia and Dystopia: Landscaping the Contemporary City*, Geografiska Annaler, Series B, *Human Geography* 84, no. 3/4, Special Issue: The Dialectics of Utopia and Dystopia (2002), 153–54.

17. MacLeod and Ward, 154. J. Allen, "Worlds Within Cities," in D. Massey, J. Allen, and S. and Pile, eds., *City Worlds* (London: Routledge, 1999): 55–97. The embedded quotation is found on page 91.

18. David Harvey, *Spaces of Hope*, 152.

19. S. Zukin, "Urban Lifestyles: Diversity and Standardization in Spaces of Consumption," *Urban Studies* 35: 830.

20. For more on this issue see MacLeod and Ward, 156.

21. Christine Boyer, "Cities for Sale: Merchandising History at South Street Seaport," 191.

22. Garret Eckbo, "Urban Nature," 224.

23. Isabelle Stengers, *In Catastrophic Times* (Open Humanities Press, 2015): 94.

24. MacLeod and Ward, 154.

25. Anthony Oliver-Smith, "What is a Disaster," "Angry Earth," 29.

26. Susanna Hoffman, "Varieties of Cultural Response," "Angry Earth," 133.

27. Susanna Hoffman, "Varieties of Cultural Response," "Angry Earth," 133.

28. Ludwig Wittgenstein, *Philosophical Investigations*, trans. G.E.M. Anscombe (London: Pearson, 1973): Section 66.

29. Garret Eckbo, "Urban Nature," *The Town Planning Review* 56, no. 2 (April 1985): 241.

30. Isabella Stengers, *In Catastrophic Times* (Open Humanities Press, 2015): 118.

31. J.L. Austin, *How to Do Things with Words*, first American edition (Oxford: Clarendon Press, 1962).

32. Maggie Nelson, *The Art of Cruelty*, 145.

33. Lauren Berlant, *Cruel Optimism* (Durham: Duke University Press, 2011): 1–2.

34. Rosemary Horrox, *The Black Death: Manchester Medieval Sources Series*, ed. and trans. Rosemary Horrox (Manchester: Manchester University Press, 1995). Horrox provides only a snippet of the original poem, the full version is provided by John W. Conlee, *Middle English Debate Poetry* (East Lansing, 1991): 51–52. I am quoting the translation provided by Horrox in her source book which she acknowledges could be rendered differently. I, however, agree with both her

translation and her motivation for rendering the language as she does in translation. The lines in question are "When thou leste wenes, venit mors te superare/ When thi grafe grenes, bonum est mortis meditari" (347). Her text also contains the material in support of my claim about Christian meditations on the value of a good death, 345–346.

Bibliography

Aaron, Jason, *The Goddamned*, Image Comics, 2017.
Athanasiou, Tom, *Divided Planet: The Ecology of Rich and Poor*, Athens: University of Georgia Press,1998.
Austin, J.L., *How to Do Things with Words*, 1st American edition, Oxford: Clarendon Press, 1962.
Baron, Naomi, "Is Reading Boring," *Huffington Post* (May 18, 2015).
Benjamin, Walter, *Illuminations*, New York: Shocken Books, 2007.
Berlant, Lauren, *Cruel Optimism*, Durham: Duke University Press, 2011.
Bhabha, Homi, *The Location of Culture*, London: Routledge Classics, 2004.
Bible, *English Standard Version*, Wheaton, IL: Crossway, 2008.
Bloom, Laura, "Why Aren't There More Female Airline Pilots? This High-Flying Woman Is Breaking Boundaries," *Forbes* (June 23, 2016).
Bocquet, Olivier, and Jean-Marc Rochette, *Snowpiercer: Terminus*, Titan Comics, 2016.
Bookchin, Murray, *The Ecology of Freedom: The Emergence and Dissolution of Hierarchy*, New York: Black Rose Books, 1991.
Brannen, Peter, *The Ends of the World: Volcanic Apocalypses, Lethal Oceans, and Our Quest to Understand Earth's Past Mass Extinctions*, New York: HarperCollins, 2017.
Brink, Susan, "A Superbug That Resisted 26 Antibiotics," *NPR* (January 17, 2017).
Brox, Ali, "'Every Age Has the Vampire It Needs': Octavia Butler's Vampiric Vision in Fledgling," *Utopian Studies* 19 no. 3 (2008).
Carrington, Damien, "Pollution responsible for quarter of deaths of young children, says WHO," *The Guardian* (March 7, 2017).
Charles, Alec, "The Meta-Utopian Metatext: The Deconstructive Dreams of Ulysses and Finnegan's Wake" *Utopian Studies* 23, no. 2 (2012).
Chen, Lei, Randall Todd, Julia Keihlbauch, Maroya Walters, Alexander Kallen, "*Notes from the Field*: Pan-Resistant New Delhi Metallo-Beta-Lactamase-Producing *Klebsiella pneumoniae*—Washoe County, Nevada, 2016," *Morbidity and Mortality Weekly Report* 66, no. 1 (2017): 33.
Chute, Hillary, *Disaster Drawn: Visual Witness, Comics, and Documentary Form*, Cambridge: The Belknap Press of Harvard University Press, 2016.
Chute, Hillary, *Outside the Box: Interviews with Contemporary Cartoonists*, Chicago: University of Chicago Press, 2014.
Cumming, Vivian, "How Many People Can Our Planet Really Support?" *BBC Earth* (March 14, 2016).
Curtis, Claire, *Postapocalyptic Fiction and the Social Contract*, New York: Lexington Books, 2012.
Damai, Puspa, "Messianic-City: Ruins, Refuge, and Hospitality in Derrida," *Discourse* 27, no. 2/3 (Spring and Fall 2005).

Deaux, George, *The Black Death*, 1347, London: Hamilton, 1969.
de Laplante, Kevin, "Making the Abstract Concrete," *Comics as Philosophy*, ed. Jeff McLaughlin, Jackson: University of Mississippi Press, 2005.
Dobrin, Sidney, and Sean Morey, *Ecosee: Image, Rhetoric, Nature*, Albany: State University Press of New York, 2009.
Eckbo, Garret, "Urban Nature," *The Town Planning Review* 56, no. 2 (April 1985).
Ellerby, Sharon, "Graphic Novels Are Gaining in Popularity," *Writing and Editing Resources* (IAPWE, November 2017).
Felski, Rita, *The Limits of Critique*, Chicago: University of Chicago Press, 2015.
Fish, Eric, "The Forgotten Legacy of the Banqiao Dam Collapse," *The Economic Observer* (February 8, 2013).
Fisher, Max, "This map shows where the world's 30 million slaves live. There are 60,000 in the U.S.," *Washington Post* (October 17, 2013).
Fox, Warwick, *A Theory of General Ethics: Human Relationships, Nature, and the Built Environment*, Cambridge: MIT Press, 2006.
Frase, Peter, *Four Futures: Life After Capitalism*, New York: Verso, 2016.
Freeman, Jacob, and John Anderies, "The Socioecology of Hunter-gatherer Territory Size," *Journal of Anthropological Archeology* 39 (2015).
Freschi, Nicholas, "Taiwan's Nuclear Dilemma," *The Diplomat* (March 14, 2018).
Ghosh, Amitav, *The Great Derangement*, Chicago: University of Chicago Press, 2016.
Glendening, Daniel, "Rucka and Lark Reunite for Dystopian Lazarus," *Comic Book Resources* (August 22, 2012).
Goodell, Jeff, *The Water Will Come: Rising Seas, Sinking Cities, and the Remaking of the Civilized World*, New York: Little, Brown, 2017.
Gottfried, Robert, *The Black Death: Natural and Human Disaster in Medieval Europe*, London: Robert Hale, 1983.
Hartogs, Jessica, "Three Years After Oil Spill, Active Clean-up Ends in Three States," *CBS News* (June 10, 2013).
Harvey, David, *Spaces of Hope*, Oakland: University of California Press, 2000.
Hoffman, Susanna, "Anthropology and the Angry Earth: An Overview," *The Angry Earth*, eds. Anthony Oliver-Smith and Susanna Hoffman, New York: Routledge, 1999.
Hoffman, Susanna, "The Worst of Times, The Best of Times: Toward a Model of Cultural Response to Disaster," *The Angry Earth*, eds. Anthony Oliver-Smith and Susanna Hoffman, New York: Routledge, 1999.
Horrox, Rosemary, *The Black Death: Manchester Medieval Sources Series*, ed. and trans. Rosemary Horrox, Manchester: Manchester University Press, 1995.
Howey, Hugh, *Wool*, Jet City Comics, 2014.
Huq, Husna, "5 Reasons Graphic Novels are the Next Big Thing at Your Library," *Christian Science Monitor* (May 6, 2013).
Huso, Deborah, "Women on the Line: Railroads still lag when it comes to hiring women, but Class I's Aim to Pick Up the Pace," *Progressive Railroading* (August 2015).
Ingraham, Christopher, "The Long Steady Decline of Literary Reading," *The Washington Post* (September 7, 2016).
Jendrysik, Mark, "Fundamental Oppositions: Utopia and the Individual," *The Individual and Utopia: A Multidisciplinary Study of Humanity and Perfection*, eds. Clint Jones and Cameron Ellis, Surrey: Ashgate, 2015.
Jones, Clint, *Ecological Reflections on Post-Capitalist Society*, Stevens Point, WI: Cornerstone Press, 2018.
Jones, Clint, *A Genealogy of Social Violence*, New York: Routledge, 2016 [Surrey: Ashgate, 2013].
Jones, Clint, *Stranger, Creature, Thing, Other: Moral Monstrosity and Our Ecostential Crisis*, Stevens Point, WI: Cornerstone Press, 2019.

Bibliography 175

Kirkman, Robert, *The Walking Dead Compendiums 1*, Image Comics, 2009.
Kirkman, Robert, *The Walking Dead Compendiums 2*, Image Comics, 2012.
Kirkman, Robert, *The Walking Dead Compendiums 3*, Image Comics, 2015.
Kirksey, Eben, *Emergent Ecologies*, Durham: Duke University Press, 2015.
Kohn, David, "The 300 Million Gallon Warning," *Mother Jones* (March/April 2002).
Langton, Lynn, "Crime Date Brief: Women in Law Enforcement 1978–2008," *Bureau of Justice Statistics* (June 2010).
Le Guin, Ursula, "Utopiyin, Utopiyang," *Utopia*, Thomas More, New York: Verso, 2016.
Legrand, Benjamin, and Jean-Marc Rochette, *Snowpiercer: The Explorers*, Titan Comics, 2014.
Lemire, Jeff, *Sweet Tooth*, vols. 1–40, Vertigo (November 2009–February 2013).
Lob, Jacques, and Jean-Marc Rochette, *Snowpiercer: The Escape*, Titan Comics, 2014.
Loewenstein, Anthony, *Disaster Capitalism: Making a Killing Out of Catastrophe*, London: Verso, 2017.
Long, Heather, "Female CEO's are at a record level in 2016, but it's still only 5%," *CNN Money* (September 29, 2016).
MacLeod, Gordon and Kevin Ward, *Spaces of Utopia and Dystopia: Landscaping the Contemporary City*, Geografiska Annaler Series B, Human Geography 84, no. 3/4, Special Issue: The Dialectics of Utopia and Dystopia (2002).
McCloud, Scott, *Understanding Comics: The Invisible Art*, New York: HarperCollins, 1994.
McKibben, Bill, *Eaarth*, New York, Times Books, 2010.
McLaughlin, Jeff, *Comics as Philosophy*, ed. Jeff McLaughlin, Jackson: University Press of Mississippi, 2005.
Mitchell, W.J.T., *Iconology: Image, Text, Ideology*, Chicago: University of Chicago Press, 1986.
Moote, A. Lloyd, and Dorothy Moote, *The Great Plague: The Story of London's Most Deadly Year*, Baltimore: Johns Hopkins University Press, 2004.
More, Thoms, *Utopia*, New York: Verso, 2016.
Morton, Timothy, *Ecology Without Nature*, Cambridge: Harvard University Press, 2007.
National Endowment for the Arts, "Results from the Annual Arts Basic Survey 2013–2015," published online.
Nelson, Maggie, *The Art of Cruelty*, New York: W.W. Norton, 2011.
Oliver-Smith, Anthony, and Susanna Hoffman, *The Angry Earth*, New York: Routledge, 1999.
Oskin, Becky, "What Would Happen if Yellowstone's Supervoclano Erupted?" *Life's Little Mysteries* (May 2, 2016).
Paik, Peter, *From Utopia to Apocalypse*, Minneapolis: University of Minnesota Press, 2010.
Parenti, Christian, *Tropic of Chaos: Climate Change and the New Geography of Violence*, New York: Nation Books, 2011.
Parker, Laura, "Ocean Trash: 5.25 Trillion Pieces and Counting, the Big Questions Remain," *National Geographic* (January 11, 2015).
Perrin, Andrew, "Who Doesn't Read Books in America?" *Pew Research Center* (March 23, 2018), published online.
Pimentel, David, and Marcia H. Pimentel, *Food, Energy, and Society*, Boca Raton: CRC Press, 2008.
Postman, Neil, *Technopoly*, New York: Alfred A. Knopf, 1992.
Prins, Gwyn, and Elizabeth Sellwood, "Global Security Problems and the Challenge to Democratic Process," *Reimagining Political Community*, eds. Daniele Archibugi, David Held, and Martin Kohler, Stanford: Stanford University Press, 1998.
Purdy, Jedediah, *After Nature: A Politics for the Anthropocene*, Cambridge: Harvard University Press, 2015.
Rangan, Daksha, "A Massive Chunk of Antarctic Ice is about to Break Off," *Science and Environment* (January 8, 2017).

Remender, Rick and Greg Tocchini, *Low*, Image Comics, 2015.
Remender, Rick and Sean Murphy, *Tokyo Ghost*, Image Comics, 2016.
Riceour, Paul, *Lectuers on Ideologies and Utopia*, ed. George Taylor, New York: Cambridge University Press, 1986.
Rodden, Robert, Floyd John, and Richard Laurino, "Exploratory Analysis of Firestorms," *Office of Civil Defense, Department of the Army* (May 1965).
Rowlands, Mark, *The Environmental Crisis: Understanding the Value of Nature*, New York: Palgrave, 2000.
Rucka, Greg, and Michael Lark, *Lazarus*, Image Comics, 2013.
Saint-Amour, Paul K., *Tense Futures*, New York: Oxford University Press, 2015.
Sloterdijk, Peter, *Bubbles*, Los Angeles: Semiotext(e), 2011.
Smith, Mick, *Against Ecological Sovereignty: Ethics, Biopolitics, and Saving the Natural World*, Minneapolis: University of Minnesota Press, 2011.
Stengers, Isabelle, *In Catastrophic Times: Resisting the Coming Barbarism*, trans. Andrew Goffey, Open Humanities Press, 2015.
Stolzenburg, William, *Where the Wild Things Were: Life, Death, and Ecological Wreckage in a Land of Vanishing Predators*, New York: Bloomsbury, 2008.
Tabachnick, Stephen, *The Cambridge Companion to the Graphic Novel*, ed. Stephen Tabachnick, New York: Cambridge University Press, 2017.
Thomson, Iain, "Deconstructing the Hero," *Comics as Philosophy* ed. Jeff McLaughlin, Jackson: University Press of Mississippi, 2005.
Tsing, Anna Lowenhaupt, *The Mushroom at the End of the World*, Princeton: Princeton University Press, 2015.
Varnum, Robin, and Christina Gibbons, *The Language of Comics: Word and Image*, Jackson: University Press of Mississippi, 2001.
Vaughn, Brian K., and Pia Guerra, *Y: The Last Man*, deluxe edition vol. 1, Vertigo, 2008.
Vaughn, Brian K., and Pia Guerra, *Y: The Last Man*, deluxe edition vol. 2, Vertigo, 2009.
Vaughn, Brian K., and Pia Guerra, *Y: The Last Man*, deluxe edition vol. 3, Vertigo, 2010.
Vaughn, Brian K., and Pia Guerra, *Y: The Last Man*, deluxe edition vol. 4, Vertigo, 2010.
Vaughn, Brian K., and Pia Guerra, *Y: The Last Man*, deluxe edition vol. 5, Vertigo, 2011.
Weisman, Alan, *The World Without Us*, New York: Picador, 2008.
Wittgenstein, Ludwig, *Philosophical Investigations*, trans. G.E.M. Anscombe, London: Pearson, 1973.
Wood, Brian, *The Massive: Black Pacific*, vol. 1, Dark Horse Comics, 2013.
Wood, Brian, *The Massive: Longship*, vol. 3, Dark Horse Comics, 2014.
Wood, Brian, *The Massive: Ninth Wave*, Dark Horse Comics, 2016.
Wood, Brian, *The Massive: Ragnarok*, vol. 5, Dark Horse Comics, 2015.
Wood, Brian, *The Massive: Sahara*, vol. 4, Dark Horse Comics, 2014.
Wood, Brian, *The Massive: Subcontinental*, vol. 2, Dark Horse Comics, 2013.
Yeh, Benjamin, "Rising Sea Levels Threaten Taiwan," *Taipei Times* (May 10, 2010).
Žižek, Slavoj, *Looking Awry*, Cambridge: MIT Press, 1992.
Zukin, S., "Urban Lifestyles: Diversity and Standardization in Spaces of Consumption," *Urban Studies* 35.

Index

Aaron, Jason 25, 27
Abbey, Edward 140
Adam and Eve 24
Africa 46, 100, 131, 133, 134, 136–138
Air Line Pilots Association (ALPA) 79
Alaska 102, 103, 119
Alexandria 65–67, 69, 70
Aliens 8, 47
Antarctica 131
anthropocentric 146
anthropogenic 7, 128, 139
apocalypse 9, 23, 35, 42, 47, 51, 53, 61, 72, 96, 99, 115, 121, 129; dystopia 29, 124, 144; graphic novels 2, 3, 7, 75, 91; social factors 11, 17, 29, 38, 39, 41, 45, 83, 84, 93, 127; utopia 10, 123
Apocalyptic 23–26
Arabian Peninsula 137
Arabian Sea 133, 134
Arizona 92, 94, 95, 141
Athanasiou, Tom 9
Atlanta, GA 49, 50–57, 60, 61, 63–66, 163n34
Atlantic Ocean 89, 99, 130
Austin, J.L. 154
Australia 75, 77, 79, 90, 98, 137

Baltimore 66, 86, 88
Banqiao Dam 141
Being-as-Belonging 146, 147
Benjamin, Walter 143
Berlant, Lauren 30, 155
Bhabha, Homi 27, 28
Black Death 26, 27, 40, 44, 160n52
Bocquet, Olivier 105, 117, 120–124, 167n8
Boston 78, 81, 87–92, 95
Boston Logan International 81, 165n28
Boyer, Christine 151
British Petroleum (BP) 97
Brooklyn, NY 77, 90

Cain 24, 25
California 40, 81, 89, 92, 95, 98, 130
capitalism 12, 16, 18, 35, 106, 145

catastrophic convergence 128
Central America 137
chemicals 46, 47, 65, 97, 111
Chernobyl 45, 46, 93
Chesapeake Bay 66
Chicago 80
China 99, 140, 141, 166n64
Chute, Hillary 3
Clinton, Hillary 33, 161n4
coal 39–41, 43, 45, 64–66, 149
Colorado 94, 103, 137, 140
communication 84, 90, 96, 132, 141
Connecticut 89, 140
Conshelf 159n23
cruel optimism 155
Cumming, Vivian 20
Curtis, Claire 10, 11
Cynthiana 49, 51, 52, 62–64, 163n34

dams 16, 140, 141, 169n46
dead bodies 44, 71, 77, 85, 86, 92, 165n28
dead zone 46, 162n20
Deepwater Horizon 97, 130
Derrida, Jacques 31, 32, 106, 167n10
diachronicity 11
disaster capitalism 16
disease 32, 36, 43–45, 76, 100, 103, 104, 111, 125, 140, 148, 166n6
disengagement 28
disentanglement 28
Disney Corporation 151
The Dispossessed 28
divine intervention 23
Donora smog cloud 21
Dulles International Airport 86
dystopia 2, 21, 22, 27, 29, 33–35, 43, 47, 103, 111, 143, 146, 148–151; disaster 15, 21; near-future 10, 15, 20, 148; (post-) apocalyptic 2, 8, 9, 11–13, 15, 19, 20, 23, 29, 31, 48, 121, 144, 145, 159n22; techno- 17, 18; worst-case scenario (disaster) 10, 15, 19, 21, 22, 25, 43, 48, 148
dystopian hope 7, 12, 71, 76, 99, 151

177

Earthjustice 64
Eckbo, Garret 149
eco-cosmopolitanism 14
Energy Justice Network 66
ensembles of selves 147
Environmental Protection Agency (EPA) 20
The Epic of Gilgamesh 26
The Eridu Genesis 26
Eve see Adam and Eve

family resemblances 153
The Flintstones 158n17
Fox, Warwick 146
France 99, 160n52, 166n64
Frankfort School 35
Franklin, H. Bruce 158n6
Frase, Peter 15
Fukushima 13, 93, 130, 132

Gaiman, Neil 6
Garden of Eden 24, 25
"gendercide" 75, 77–88, 92–94, 96, 98, 99, 166n64
Georgia 49, 51, 54–57, 60, 61, 64
Gibbon, Edward 7
The Goddamned 25
graphic novels 1–8, 14, 15, 23, 142, 144
Great Basin 137
"great flood" 24, 25, 160n50
Guerra, Pia 75, 77, 78, 81
Gulf of Mexico 40, 97

Harvey, David 149, 150
Hoffman, Susanna 4, 13, 30, 155
Hoover Dam 140
Howey, Hugh 19, 21, 159n22
H2OME 159n26
Huxlian 18
hydraulic fracking 40, 41

India 13, 26, 37, 134, 137, 153
International Society of Women Airline Pilots 79
invasive species 47
Israel 75, 78, 79, 88, 89, 94, 99

Jameson, Fredric 158n6
Japan 18, 96, 98, 99, 132, 153, 166n61, 166n64
Judeo-Christian-Islamic 23

Kansas 81, 92, 94, 166n59
Kirkman, Robert 49, 51, 58, 60–62, 64, 67, 70–72, 74–76, 100, 102–104, 133, 135, 145
Kirksey, Eben 147
Klee, Paul 7
Klein, Naomi 16

Lacanian 4, 9
LAX 81
Lazarus 15–17, 35
Le Corbusier 149
Legrand, Benjamin 111, 113, 115–118, 120, 122, 123, 167n8
Le Guin, Ursula 27, 28
liquid natural gas (LNG) 97, 163n33, 165n54
Lob, Jacques 105, 106, 110, 111, 113, 117, 120–124, 126, 167n8, 167n9
The Location of Culture 27
Loewenstein, Anthony 16
Looking Awry 9
Low 19, 20

Marcuse, Herbert 35
Marx, Karl 35, 146
Maryland 64, 86
Marysville, OH 91, 92
McCloud, Scott 2, 3, 115
McKibben, Bill 144, 145
medicine 37, 43, 44, 84
messianic 33, 144
Miller, Frank 6
Minnesota 103
A Modern Utopia 34
Moore, Alan 6
mortality rate 36, 160n52
Morton, Timothy 144, 146

National Endowment for the Arts 5
Nebraska 94, 102
Nelson, Maggie 26
New York City 75
Noah 24–27, 121
North America 26, 57, 137, 169n29
nuclear 13, 36, 37, 45, 46, 65, 66, 74, 78, 93, 94, 100, 101, 104–111, 115, 117, 121–123, 130, 132, 139–142; summer 125; winter 105, 107–110, 115, 121, 122, 125, 139

Occupy Movement 17
O'Hare International Airport 80
oil 37, 39, 40, 41, 43, 45, 51, 65, 83, 97, 130, 133–135, 169n29
Orwellian 17

Pacific Ocean 26, 93, 97, 99, 141
Parenti, Christian 128, 145
Paris, France 99
Pew Research Center 5
Plato 160n50
political ecology 14, 15, 152
Poseidon Resort 159n26
post-apocalyptic 2, 4, 6, 8–14, 19, 20, 22–26, 35–48, 55, 69, 71–74, 76, 86, 88, 90–97,

102–106, 111, 112, 115, 121, 126, 127, 135, 136, 138–142, 144, 145, 151, 153
post-human 103, 104, 114, 122
Potomac River 66

Reagan, Ronald 86
refugee crisis 44, 45, 133, 134, 137, 169n37
Remender, Rick 15–20
responsive cohesion 147
Ricoeur, Paul 34
Rocky Mountains 103
Rucka, Greg 15–17

Sagan, Carl 105
Saint-Amour, Paul 106
Sanders, Bernie 33
Saudi Arabia 138
SEALAB 159n23
slavery 139, 141
Sloterdijk, Peter 74
Snow Crash 159n27
social contract 11
Somalia 134, 137
South Carolina 57, 64
speciesist 146
Star Trek 112
Stengers, Isabelle 154
storms 42
Sub-Biosphere 159n23
"superbug" 44
supervolcano 108, 109
synergy 22, 24, 36, 66, 141, 145

tabula rasa 13, 143
Taiwan 131, 132
Tokyo Ghost 15, 16, 18, 35
The Tower of Babel 26
Trump, Donald 33

United States 13, 14, 21, 36, 39, 40, 42–48, 57, 64, 66, 69, 72, 78–80, 82, 84, 93, 98, 103, 108, 128, 140, 141, 145, 149, 161n55
utopia 10, 11, 17, 27, 28–35, 48, 72, 91, 122, 143, 145, 148, 149–151; and erasure 29, 30, 31; of ruins 31, 48

Vaughn, Brian K. 75–79, 81–85, 87–90, 92, 95, 96, 98–100, 102–104, 133, 135

The Walking Dead 75, 76, 100, 102, 104, 130, 139, 154
Washington D.C. 49, 61, 63, 65, 66, 68, 75, 85, 88, 90, 93
Wells, H.G. 34, 35
Wittgenstein, Ludwig 153
Wood, Brian 127
Wool 19–21
World Health Organization 21, 161n55
world population 19

Y: The Last Man 102, 130

Zizek, Slavoj 9

www.ingramcontent.com/pod-product-compliance
Ingram Content Group UK Ltd.
Pitfield, Milton Keynes, MK11 3LW, UK
UKHW042014140426
5217IPUK00015B/1172